SOVIET
CONQUEST
FROM
SPACE

SOVIET CONQUEST FROM SPACE

PETER N. JAMES

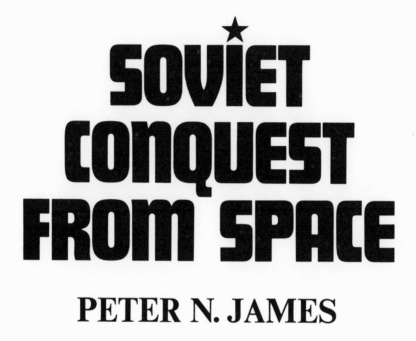

ARLINGTON HOUSE·PUBLISHERS

NEW ROCHELLE, N. Y.

Manufactured in the United States of America

Library of Congress Cataloging in Publication Data

James, Peter N   1940-
   Soviet conquest from space.

   Bibliography: p.
   1. Aeronautics, Military--Russia. 2. Astronautics
--Russia. 3. Russia--Military policy. I. Title.
UG635.R9J35   358'.8'0947   73-22457
ISBN 0-87000-224-4

To my loving wife
Diane

Contents

Illustrations

Tables

Acknowledgments

I wish to express my appreciation to personnel within the U.S. political-academic-military-industrial complex for providing me with weekly voluminous documents, reports, and background material pertinent to the subject of this book. I wish particularly to thank (1) Senator Clinton P. Anderson, New Mexico (retired); (2) the late Senator Allen J. Ellender, Louisiana; (3) Senator Henry M. Jackson, Washington; (4) Senator Walter F. Mondale, Minnesota; (5) Senator William Proxmire, Wisconsin; (6) Congressman Paul G. Rogers, Florida; (7) General Jacob E. Smart, assistant administrator for DOD and interagency affairs, National Aeronautics and Space Administration; (8) Richard T. Mittauer, director of public information, National Aeronautics and Space Administration; (9) Helmut Dederra, Messerschmitt-Bolkow-Blohm, Ottobrunn, West Germany; (10) Sam F. Iacobellis and E. F. Hogan, Rocketdyne, Division of Rockwell International Corporation (formerly called North American Rockwell Corporation), Canoga Park, California; (11) Richard B. Phillips, Jet Propulsion Laboratory, Pasadena, California; (12) Dr. Richard Pipes, director, Russian Research Center, Harvard University; and (13) John Polasek, Jr., product information manager, Grumman Aerospace Corporation, Bethpage, New York.

For obvious reasons, the sources of some of the information within cannot be disclosed. And finally, I wish to thank my neighbor, John D. MacArthur, of the Colonnades Beach Hotel, for employing me while I worked on the manuscript; David Franke, senior editor of Arlington House, for working with me on this book; my brother Herb, of Weeber-James Art Associates in Cleveland, Ohio, for illustrations of the Russian space systems; and my wife, Diane, for typing the draft and final manuscript and for being at my side at each international conference—words cannot express her contribution to this effort.

Peter N. James

Foreword

It is the primary objective of *Soviet Conquest From Space* to inform the public of matters that have been unjustifiably withheld from them because of political rather than security reasons. *Soviet Conquest From Space* draws heavily on my experiences with personnel affiliated with the following organizations:

UNITED STATES	*SOVIET UNION*
U.S. Central Intelligence Agency	The Committee for State Security of the Soviet Union
U.S. Defense Intelligence Agency	The Chief Intelligence Directorate of the General Staff
U.S. Department of Defense	Ministry of Defense of the Soviet Union
National Aeronautics and Space Administration	The State Committee for Science and Technology
USAF Foreign Technology Division	Academy of Sciences of the Soviet Union

| U.S. Department of State | The Commission for the Exploration and Use of Outer Space |
| Numerous U.S. institutes and corporations | Numerous U.S.S.R. institutes and facilities |

Private records show that I assisted the CIA in exploiting Soviet science and technology during 1965–1972 and the United States Air Force Foreign Technology Division (FTD) during 1968–1971, when I was manager of the Foreign Technology Program at Pratt & Whitney's Florida Research and Development Center. During 1965 through 1970 I attended international scientific conferences attended by Soviet scientists and intelligence officers in the following cities:

> Athens, Greece
> Madrid, Spain
> Belgrade, Yugoslavia
> Venice, Italy
> Paris, France
> Mar del Plata, Argentina
> Rome, Italy
> Konstanz, Germany

At these conferences I conversed and frequently held private meetings with (1) Soviet scientists and engineers responsible for the development of space and defense weapons systems, (2) Soviet espionage agents responsible for exploiting U.S. scientists, engineers and technology to the advantage of the Soviet Union, and (3) officials responsible for managing the Soviet military-industrial-political complex. On one hand I collected intelligence information through personal contact with the Soviets, while on the other, I analyzed intelligence at the Florida Research and Development Center. This experience was unique because the U.S. intelligence community (i.e., Central Intelligence Agency, Defense Intelligence Agency, et al.,) is rigidly compartmentalized into collectors of intelligence (spooks) and analyzers of intelligence (analysts); a spook is not permitted to analyze the information that he collects, and an analyst is not permitted to have personal contact with members of the Eastern bloc to collect intelligence. I did both with excellent results.

After I wrote my second assessment of the Soviet space program and their military-industrial-political complex, an 800-page report classified "Secret—No Foreign Dissemination" and distributed in late 1970 to the Executive Office of the President, the Central Intelligence Agency, the Department of Defense, the National Aeronautics and Space Administration, and other organizations, I became involved in a heated dispute with Pratt & Whitney Aircraft management and intelligence officers from the Air Force Foreign Technology Division over the prospects of writing an intelligence assessment on the Soviets for public consumption. It was my position that certain information concerning the Soviet espionage establishment and their military-industrial-political complex belonged in the public domain, and it was not in the interest of the United States to withhold the information from the public, though it might be embarrassing to various U.S. government officials and other parties. In December 1970, Pratt & Whitney Aircraft management bluntly told me that if I attempted to write such a book I would be fired. Other officials told me that the public was "too dumb" to understand the subject intelligently.

In 1971, I secretly worked on the book and was granted a leave of absence without pay for the summer. During this leave, I attended international conferences in Dubrovnik, Yugoslavia; Marseilles, France; and Brussels, Belgium to obtain more material for this book. At these conferences I continued to hold secret meetings with Russians in restaurants, lounges, passenger ferryboats, and various hotel rooms. On October 4, 1971—my first day back to work—I was fired by Pratt & Whitney when I voluntarily disclosed that I attended three international meetings to research material for this book.

In short, the material presented within is based on my background of approximately nine years as a space systems analyst, seven years of association with Soviet scientists, engineers, managers, and intelligence agents, and author of two secret reports and private memorandums on Soviet rocket and space technology. I was frequently asked to critique the latest intelligence assessments of the Air Force Foreign Technology Division at Wright-Patterson Air Force Base (near Dayton, Ohio) and the Defense Intelligence Agency in the areas of Soviet rocket and space technology and the organization of the Soviet military-industrial-political complex. As is shown in later sections, I was the first intelligence analyst to document, among other things, the existence of an aggressive military-oriented Soviet space shuttle development program. It is with this background that I undertook the project of writing an intelligence assessment on the Soviets for the American who does not have a technical or scientific background, but would basically like to know—without any doubletalk—what is going on behind the Iron Curtain.

Note to the Reader

I have taken the liberty to present material in this book in the same format that I used in the 800-page secret intelligence report that I prepared for Pratt & Whitney Aircraft. In that report the significant conclusions were boxed in, as shown for this paragraph, and were identified by Subjective Analysis Summaries (SAS) numbers located in the lower left-hand corner of each box. The SAS sections are referenced in later parts of the book; page 243 contains an SAS page index for your benefit. After reading *Soviet Conquest From Space*, you will find that the SAS sections are convenient for refreshing your memory at a later date without having to reread the main text of the book.

SAS NO. ___

I have also included excerpts from my personal notes on Soviet space and weapons systems, intelligence officers, and personnel. These notes are classified "Confidential or Secret—No foreign Dissemination" in U.S. intelligence files and were privately transmitted to personnel affiliated with the U.S. intelligence community during 1965–1971. Some intelligence information presented within has never

V. I. Lenin

L. I. Brezhnev

A. N. Kosygin

PLATE 1   The Soviets

BLAGONRAVOV   DIMENTIEV   DORODNITSYN   GLUSHKO

GRODZOVSKIY   KHODAREV   KELDYSH   KIRILLIN

KOROLEV   B. PETROV   G. PETROV   SEDOV

SHATALOV   SOKOLOVSKY   SYCHEV   YELISEYEV

PLATE 2   The Soviets (continued)

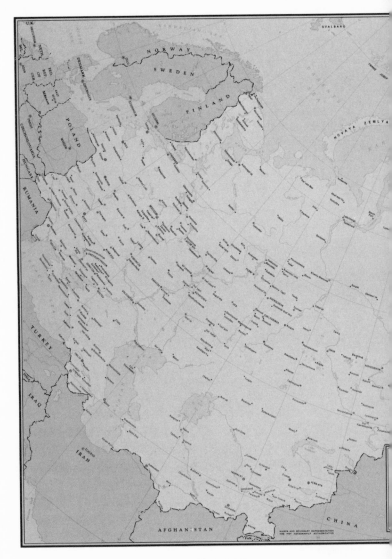

PLATE 3

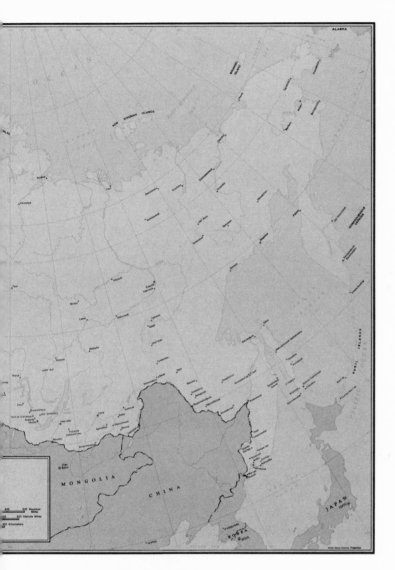

The Soviet Union

been released to the U.S. intelligence community and is presented here for the first time. I have made every attempt to provide you with information so that you can challenge the competence and wisdom of persons who rationalize their behavior on the basis that "the public is too dumb to understand the workings of the government" or "the details should be left to the people in Washington because that is their job." Remember: numerous U.S. government officials have jobs today because they have successfully brainwashed the public into believing that some government programs and events behind the Iron Curtain are too complex and technical to understand. If certain government officials become upset over this book (one government man used the word "alarmed" when he heard that I was writing it) it is because they expect to be questioned and judged by their superiors and a formerly docile and uninformed public.

Though I have criticized certain branches of the U.S. government and other organizations, *Soviet Conquest From Space* is not intended to, nor should it jeopardize the security of the United States or its allies. I hope that this work contributes to a better understanding of the freedoms we currently enjoy, those matters that affect the security of the United States and the international community, and a more wholesome environment where some government officials have as much regard for the American citizen and the national defense of the United States as they have for themselves.

Show your identification badge to the staunch-faced security officer standing in the doorway to the briefing room (your driver's license will do). Be sure to sign in at the registration desk and list the highest level Department of Defense (DOD) security clearance that you currently hold (if you have no DOD security clearance, then you are invited as my guest), and find a seat in the briefing room, perhaps next to the gray-haired Air Force colonel smoking a cigar, or behind the middle-aged partially bald, black-haired man with the dark glasses, who looks like he could be an intelligence officer called in from the field. The briefing starts on page 29 as soon as you find your seat and is written in language that everyone understands.

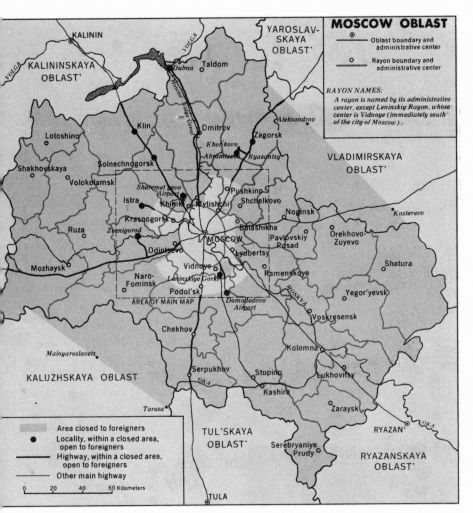

PLATE 4 Moscow and Vicinity

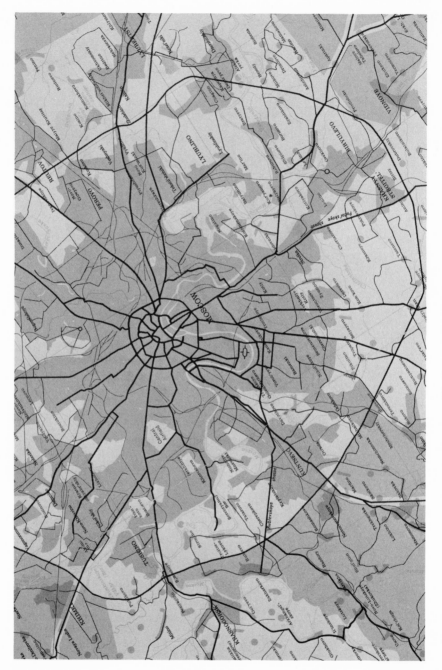

PLATE 5 Moscow

28

Chapter 1

The Rulers

The Soviet leadership believes that a war is a continuation of a policy by another means; the essence of war as a continuation of politics does not change with changing technology, armaments, or the development of nuclear weapons; nuclear war is acceptable as another means of furthering the policies of the Communist Party of the Soviet Union.

The trends shown in General Bernard Schriever's July 1967 report to the House Armed Services Committee revealing a significant deliverable nuclear megatonnage advantage for the Soviets over the United States during the early 1970s are valid and cannot be justifiably challenged.

While some intelligence analysts have questioned the logic of developing a massive nuclear overkill capability, the data shows that the Soviets do not share their view.

The current Soviet ICBM force represents a threat to the U.S. Minuteman ICBM force; the Soviets have rejected the concept of strategic-nuclear parity and are striving for absolute strategic superiority in accordance with the guidelines dictated by the Communist Party of the Soviet Union.

Soviet ICBM silos have been secretly designed with a reload capability. Intelligence estimates of the number of operational Soviet ICBMs (prior to the Moscow Summit of May 1972) do not take into account this reload factor.

It is the Soviet number one objective to achieve military superiority in space through the aggressive development of the following three systems: (1) a network of orbiting space stations, (2) a fleet of low-cost manned and unmanned reusable earth orbital spacecraft systems, and (3) a fleet of low-cost reusable space shuttle systems to provide earth-to-orbit logistics support for the planned spacecraft-space station complex.

The U.S.S.R. has been engaged in an aggressive classified military-oriented space shuttle development program since 1967. This program is managed by the Moscow Aviation Institute and the Central Aerohydrodynamics Institute. The Soviet shuttle will be capable of delivering considerably more payload into earth orbit than the U.S. space shuttle system. The Soviets lead the United States in the development of a space shuttle system by two to three years and have already flight-tested an unpowered prototype of the Rocketoplan (the upper stage of the Soviet space shuttle).

To support their military manned and unmanned maneuverable spacecraft programs, the Soviets simulated a space refueling operation by secretly transferring nitric acid and kerosene propellants during an unmanned spaceflight mission during 1969 or before. The Soviets currently lead the United States in the development of manned orbital spacecraft by about five years, and U.S. plans for the development of orbital spacecraft systems during the space shuttle era (late 1970s, 1980s) have been deferred because of budgetary reasons.

The Soviets lead the United States in the development of earth orbital space stations by three to five years, and U.S. plans to develop a space station beyond Skylab (1973) have been deferred because of budgetary reasons.

The Soviets are developing a lethal laser weapon system for use in orbital spacecraft and space stations. Significant Soviet laser work is conducted in a secret laser facility built during the 1960s in the steppes of Siberia.

It is the Soviet objective to develop an orbiting defense network that can neutralize U.S. spacecraft and space stations, ICBMs and ABMs. If the Soviets continue on their present course and the United States executes its space and defense plans as outlined by the Secretary of Defense and the head of the National Aeronautics and Space Administration, it must be concluded without reservation that the Soviets will achieve clear-cut military, strategic, and space superiority over the United States.

Some of the aforementioned conclusions are documented in the 800-page secret report and other private memorandums that I authored on the Soviet space and defense establishments; they have been subsequently corroborated by reliable sources. Currently, special groups within the U.S. intelligence community keep abreast of Soviet developments in the critical areas cited. However, amidst private and classified rhetoric, a determined effort has been made to suppress some of these developments from the public because of political reasons. The political sensitivity of the problem surfaces when one asks: how is it possible for the United States to spend approximately 80 billion dollars a year on defense and find its national security gravely threatened by a nation (i.e., the Soviet Union) that has a gross national product that is about one-half of ours? Who is responsible for this mess? These questions can only be answered if one has an understanding of how the Soviets conduct their business on space and defense matters, what their secret plans are for the coming decades, and what their real capability is—the subject of this book.

One of the most sobering, but hard-to-swallow facts that arose during the preparation of recent intelligence assessments on the Soviets was that the U.S. intelligence community is much too compartmentalized, and in many instances its personnel are too specialized to provide reliable assessments of the Soviet space and defense establishments. For example, the Soviets achieved their current competitive and powerful position in the world because of their space and weapons systems inventory, and this could be attributed to their closely-knit political-espionage-academic-military-industrial complex. Yet the current organization of the U.S. intelligence community precludes numerous qualified government analysts from conducting an integrated or systems assessment of the Soviet military and strategic capability, their science and technology, and their space and defense weapons systems; Soviet work in some specialized areas is not being related to significant work in other areas. I confirmed that this was the case through private discussions with personnel within the U.S. intelligence community and the review of hundreds of finished intelligence reports prepared by the Central Intelligence Agency (CIA), the Defense Intelligence Agency (DIA), and the Air Force Foreign Technology Division (FTD).

The paragraphs below discuss the Soviet political-espionage-academic-military-industrial complex and provide the basic groundwork for conducting a systems assessment on the Soviets. Though the material is written for the layman, and it only scratches the surface, I hope that it encourages U.S. government intelligence analysts to relate their specialized work to other significant Soviet developments, for this must be done if our leaders are to have a better understanding of the Soviets and their motives.

On paper the Soviet government consists of three branches:

(1) *The Communist Party of the Soviet Union* (CPSU), headed by General Secretary (First Secretary) Leonid I. Brezhnev, formulates policy within the Soviet Union, even though only 14.5 million Soviets out of a total population of about 250 million (i.e., less than 6 percent)

belong to the Communist Party. The CPSU is often referred to as "the Party."

(2) *The Council of Ministers of the Soviet Union*, headed by Premier Aleksei N. Kosygin, executes the policies of the Communist Party. The Council of Ministers is sometimes referred to as "the Government" or the Executive branch.

(3) *The Supreme Soviet*, headed by President Nikolai V. Podgorny, ratifies and approves the policies of the Communist Party and the Council of Ministers. The Supreme Soviet is sometimes referred to as "the Legislature" or "the Parliament."

In reality, the General Secretary (Leonid Brezhnev) holds the power in the Soviet Union, followed by the Premier (Aleksei Kosygin) and the ceremonial President of the Supreme Soviet (Nikolai Podgorny), whose branch provides the rubber stamp approval of the policies dictated by the Communist Party.

Plate 6 is a simplified organizational chart of the Soviet government. Attention should be placed on the subservient role of the Council of Ministers and the Supreme Soviet to the Politburo of the Communist Party, which is discussed below.

The sixteen most powerful men in the Soviet Union belong to a body called the Politburo (Politbureau); these men are elected by the Central Committee of the Communist Party. The Central Committee, a body of several hundred high-level Communist Party members, is elected by the Congress of the Communist Party, which meets approximately every five years. The 24th Congress of the Communist Party was held in Moscow's Palace of the Congresses during March 30 through April 9, 1971. At this Congress, 4,949 Communist delegates, representing 14,455,321 Communist Party members throughout the Soviet Union, elected a Central Committee made up predominantly of Brezhnev men.

The key to political power in the Soviet Union is to have the support of both the Politburo and Central Committee. Brezhnev solidified his support in the Politburo during the 24th Congress when the Central Committee voted to expand the ruling (voting) membership of the Politburo from eleven to fifteen men; the new additions to the Politburo have been generally regarded as Brezhnev's proteges. Then, during April 1973, the Politburo was expanded to sixteen men, and two of Brezhnev's rivals on the Politburo were ousted and demoted to retirement. Under the Soviet system, the Politburo, though it is supposedly an elected body, is a self-perpetuating political machine that exerts tremendous influence over the membership of the party's Central Committee and Congress by effective use of (1) political

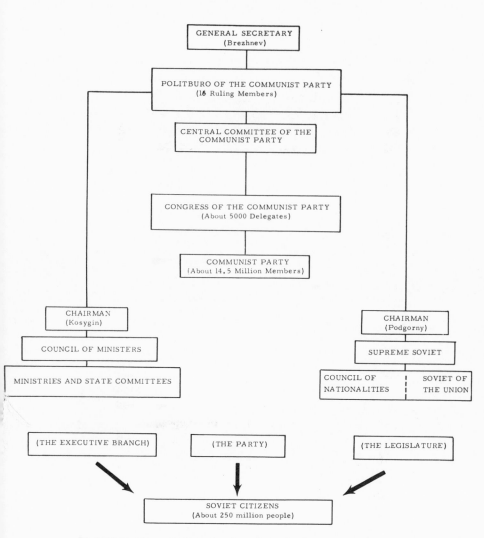

PLATE 6 **General Organization of the Soviet Government**
SAS-1

payoffs and (2) the Committee for State Security of the Council of Ministers (i.e., the KGB, Komitet Gosudarstvennoi Bezopasnosti, which is the all-powerful and pervasive secret police and intelligence service in the Soviet Union, discussed in later paragraphs).

Table I lists the sixteen full or voting members of the Politburo, including the four new members who were elected during the 24th Party Congress in 1971, and the changes made during April 1973. Also listed are candidate or non-voting members of the Politburo. Candidate members hold key positions in the Soviet government and their advice is often sought by the sixteen full members, but they have no vote on issues decided by the Politburo.

The key to achieving absolute dictatorial power in the Soviet Union is to simultaneously become the head of the Communist Party (General Secretary) and the Council of Ministers (Premier). Table II lists the Soviet leaders during the past two decades and shows that Stalin and Khrushchev achieved dictatorial status during the Soviet's turbulent history, when they were solely responsible for formulating policy (the Communist Party) and executing that policy (Council of Ministers).

A Soviet leader can become the head of both the Communist Party and the Council of Ministers when he has the overwhelming support of the members of the Politburo, Central Committee, and the Supreme Soviet. Currently, General Secretary Leonid Brezhnev has more support in these bodies than he ever had, and he has succeeded in diminishing the power of some of his archrivals. If a Soviet leader

SOVIET POLITBURO

FULL (VOTING) MEMBERS	CANDIDATE (NON-VOTING) MEMBERS
BREZHNEV, LEONID KOSYGIN, ALEKSEI PODGORNY, NIKOLAI SUSLOV, MIKHAIL KIRILENKO, ANDREI PELSHE, ARVID MAZUROV, KIRILL POLYANSKY, DMITRY SHELEPIN, ALEXANDER	DEMICHEV, PYOTR MASHEROV, PYOTR MZHAVANADZE, VASILY RASHIDOV, SHARAF USTINOV, DMITRY ROMANOV, GRIGORY
GRISHIN, VIKTOR KUNAYEV, DINMUHAMED SHCHERBITSKY, VLADIMIR KULAKOV, FEDOR	NEW MEMBERS April 1971
GROMYKO, ANDREI GRECHKO, ANDREI ANDROPOV, YURI	NEW MEMBERS April 1973

The Politburo is controlled by predominately Brezhnev men. During April 1971 and April 1973 the Plenary Meeting of the Central Committee strengthened Brezhnev's position by electing seven new full members to the Politburo and ousting Pyotr Shelest and Gennadi Voronov -- two of Brezhnev's rivals. This trend is expected to continue.

TABLE I Soviet Politburo
SAS-2

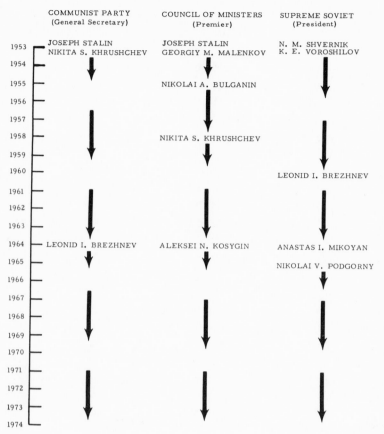

	COMMUNIST PARTY (General Secretary)	COUNCIL OF MINISTERS (Premier)	SUPREME SOVIET (President)
1953	JOSEPH STALIN NIKITA S. KHRUSHCHEV	JOSEPH STALIN GEORGIY M. MALENKOV	N. M. SHVERNIK K. E. VOROSHILOV
1954			
1955		NIKOLAI A. BULGANIN	
1956			
1957			
1958		NIKITA S. KHRUSHCHEV	
1959			
1960			LEONID I. BREZHNEV
1961			
1962			
1963			
1964	LEONID I. BREZHNEV	ALEKSEI N. KOSYGIN	ANASTAS I. MIKOYAN
1965			NIKOLAI V. PODGORNY
1966			
1967			
1968			
1969			
1970			
1971			
1972			
1973			
1974			

TABLE II Soviet Leadership Since Stalin

assumes a dictatorial position, it can be expected that he will remain in that position until his death (e.g., Stalin's death in 1953), or because his policies clearly were against the interests of the Soviet state and members of the Politburo were willing to risk a showdown vote before the Central Committee, challenging the leadership of the General Secretary (e.g., the fall of Khrushchev in 1964). There is no room for error, however: If the Central Committee upholds the leadership of the General Secretary, as was the case in 1957 when the Politburo attempted to unseat Khrushchev and failed, the General Secretary then has the necessary Central Committee support to purge the Politburo of the dissenters and assign them to menial party assignments. During the Stalin era the purge victims were usually treated brutally and imprisonment or death was the final solution. If the Central Committee is successful in removing the General Secretary, as was done to Khrushchev in 1964, it can be expected that another member of the Politburo will assume the General Secretary's position and the ousted General Secretary will be permitted to lead a private life far removed from the Soviet political arena.

Many members of the Politburo shown in Table I moved up through the Soviet political hierarchy during the Stalin era, when force and violence was commonly used to suppress or eliminate opposition; the current Soviet leaders, Brezhnev and Kosygin, belong in this category. Other members of the Politburo such as Podgorny, Suslov, and Pelshe served as Stalin's hatchet men during the bloody purges of the 1930s and 1940s. Kirilenko, who has become active in Soviet Eastern bloc affairs, and Polyansky, a farm and agricultural specialist, also began their careers under Stalin in the 1930s. Shelest, another hardline Communist, whose party membership dates back to the late 1920s, was responsible for the Ukrainian crackdown on dissidents and intellectuals. Shelest also opposed the Moscow Summit Conference of May 1972 and was ousted during the Politburo purge of April 1973. The youngest member of the Politburo, Shelepin, was formerly head of the KGB, which is known for its police-state oppression and surveillance of Soviet citizens and party members, and espionage activities against the United States. Shelest and Shelepin are Brezhnev's rivals.

In general, the Soviet Union is controlled by an elite clique of experienced hard-line Communists who began their careers during the Stalin era. The significant point worth noting is that the Politburo has resorted to the ruthlessness of the Stalin era to perpetuate its rule and solve modern-day difficulties with foreign countries, as demonstrated by the Soviet invasions of Hungary in 1956 and Czechoslovakia in 1968; the personal ambitions of this handful of men are limited only by the capabilities of the espionage and academic-military-industrial establishments that are at their disposal. The sections that follow discuss these establishments.

SAS-3

Chapter 2

The Soviet
Espionage Establishment

The Soviet espionage establishment consists of two very active organizations: (1) The KGB (Komitet Gosudarstvennoi Bezopasnosti), which is known as the Committee for State Security, and (2) the GRU (Glavnoye Razvedyvatelnoye Upravlenie), which is known as the Chief Intelligence Directorate of the General Staff of the Ministry of Defense. On paper, both intelligence organizations report to Premier Aleksei Kosygin's Council of Ministers, but in reality their orders originate from the Central Committee of the Communist Party.

The KGB is the more powerful and influential of the two intelligence organizations and is responsible for maintaining internal security—its secret police even watch Communist Party and GRU personnel. The KGB conducts espionage operations against the West and specifically the United States. In relation to U.S. organizations the KGB can be thought of as being a combination of the Central Intelligence Agency (CIA), the Federal Bureau of Investigation (FBI), and the Secret Service. The head of the KGB as of this writing is Yuri Vladimirovich Andropov, who was the Soviet ambassador to Hungary during the ill-fated Hungarian Revolution in 1956, and is currently a full member of the Politburo (refer to Table I).

GRU espionage operations against Western bloc nations parallel those of the KGB, and the two intelligence organizations often compete against each other, but many GRU operations must be approved by the KGB since the KGB is responsible for conducting background investigations and issuing security clearances on GRU personnel. In some ways the GRU can be compared to the U.S. military intelligence services, and particularly the Defense Intelligence Agency (DIA); it is mainly concerned with scientific, technical, tactical, and strategic intelligence. By order from the Central Committee, both the KGB and GRU have been instructed to conduct espionage operations against the United States to exploit or subvert our science, technology, economy, and political system to the advantage of the Communist Party.

As is the case with the CIA, the Soviet espionage apparatus can be categorized into collectors of intelligence (spooks) and analyzers of intelligence (analysts). Both the KGB and GRU have massive espionage networks throughout the world, made up of clandestine operators who have assumed new identities and have infiltrated into the fabric of our society. These spies are called illegals because they usually have entered the United States illegally with false passports; they rely on clandestine means of communications with Moscow, such as wireless radio and dead drops—hiding places that permit agents to transmit messages and money to each other without being in physical contact. The best known modern-day illegal apprehended by the FBI was alias Emil Goldfus, who was known by his friends in New York as a photographer, artist, and jewelsmith, but was in fact KGB Colonel Rudolph Ivanovich Abel, who headed a spy ring in North America. Colonel Abel was the spy who was exchanged for U-2 pilot Francis Gary Powers in 1962.

The bulk of Soviet espionage committed in the United States is coordinated by intelligence officers who are assigned to foreign embassies, trade missions, tourist agencies, government newspaper offices and the United Nations. Washington and New York are the major centers of espionage activities against the United States, and the FBI has special counterintelligence officers in these cities. The work of Soviet intelligence officers operating in the states is coordinated by the KGB and GRU in Moscow, with the intelligence agencies of the Eastern bloc satellite countries coordinating their activities with Moscow as well.

Intelligence officers, who have cover occupations—such as the scientific attache to an embassy, editor of a foreign news publication, or translator at the United Nations—are known as legals; they are legally working for their government, though in a cover assignment.

Soviet embassies are known to be staffed with a high percentage of espionage agents; the late Colonel Oleg Penkovskiy, the high-ranking GRU intelligence officer who voluntarily transmitted the innermost Kremlin secrets to Western intelligence agents during the early 1960s before he was apprehended by the KGB and executed, stated that approximately 60 percent of Soviet embassy personnel were intelligence officers, with KGB officials outnumbering their GRU counterparts by about two to one. A significant point that Penkovskiy made was that the Soviets are involved in espionage operations against the West on such a grand scale that Westerners could not comprehend the magnitude of the Soviet effort. The legals known to U.S. counterintelligence agents, primarily the FBI, operate out of the Soviet embassy on Sixteenth Street, N.W., in Washington; Amtorg, the Soviet trade agency; Aeroflot, the Soviet national airline; Tass, the Soviet press agency; and the Soviet Mission at the United Nations in New York. Soviet legals have consistently had problems with Western governments and counterintelligence forces.

During September 1971, the British government, fed up with the Soviet legals' espionage activities conducted from the Russian embassy in London, expelled 105 spies who operated from the embassy under diplomatic cover. The British dramatized the expulsion over television by showing films of a Soviet "diplomat" picking up intelligence information at a dead drop in the serene English countryside.

In the United States in February 1972, the FBI arrested Valery Markelov, a Russian translator at the United Nations, because he attempted to acquire classified documents showing the design characteristics of the Navy's advanced fighter aircraft, the F-14A, from an aerospace engineer employed by the Grumman Corporation. These charges, however, were dropped prior to President Nixon's trip to Moscow. And in August 1972, a military court convicted Master Sgt. Walter T. Perkins for trying to pass vital U.S. defense secrets to Soviet intelligence officers affiliated with the Russian embassy (legals) in Mexico City.

The significance and merits of Soviet espionage against the United States were implicitly revealed in Marshal of the Soviet Union V. D. Sokolovsky's *Military Strategy*, a most significant document published by the Military Publishing House in Moscow in 1968. This document showed that Soviet military strategy is predicated on an in-depth understanding of our social-economic-political-military-industrial complex. To comply with KGB intelligence requirements, the necessary intelligence information can only be obtained by the meticulous and fanatical accumulation of the most minute details of our society by thousands of trained and dedicated Communists and Communist sympathizers in and outside the United States. The Central Committee of the Communist Party of the Soviet Union has always ensured that the KGB and GRU are provided with adequate funds for executing this most arduous task because they firmly believe that communism can only triumph in the United States through revolution or the use of external force to overthrow our government; the prerequisite for this movement during peace or war is the understanding of our strengths and weaknesses, and this can only be done through a massive intelligence exploitation program directed against the United States, where no stone is left unturned.

The Soviet espionage program against the United States is so massive that many intelligence requirements dictated by the Central Committee seem both ridiculous and innocuous by American standards. For example, Soviet espionage agents, legals and illegals, routinely photograph U.S. government buildings, railroads, airports, and ports, and these photographs are eventually filed in the KGB Headquarters in Dzerzhinsky Square in Moscow. Yet an American observing someone with a camera near these facilities would not give the party a second thought. However, if that American were to photograph similar facilities in the Soviet Union, there is a very good possibility that he would be apprehended within seconds and subsequently charged with espionage by the KGB. The reason is simple: intelligence specialists from both the Soviet Union and the United States recognize the value of photographs of this nature for (1) preparing intelligence estimates, (2) equipping an illegal with forehand information prior to his entry into the United States, (3) briefing saboteurs of likely targets in the event hostilities should break out between the superpowers, and (4) pinpointing tactical and strategic targets. Because ours is an open society we concede this advantage to the Soviets to preserve our individual freedoms. Again, Eastern bloc legals and illegals frequently visit U.S. libraries and learn about our strengths and weaknesses from our publications, particularly technical magazines that describe our weapons systems in such detail that upwards of 90 percent of the useful technical intelligence about the United States is acquired by the Soviets through these publications. Libraries are also used by spies to obtain addresses and background information on living and dead persons. This information is eventually used as cover stories for illegals who are currently being trained in the best Soviet spy schools to infiltrate our society. Then again, no

American would be allowed to routinely visit Russian libraries to acquire similar information; travel is rigidly controlled in most parts of the Soviet Union.

Because the Soviets believe that they must surpass the United States in science and technology to achieve military and strategic superiority, the KGB and GRU have gone to great lengths to acquire the West's secrets, especially on space and defense weapons systems. Their emphasis on science and technology dates back to the teachings of Vladimir Lenin, and it came to fruition when they launched Sputnik on October 4, 1957. Sputnik not only signalled the beginning of the space age but it raised Soviet esteem in the eyes of the world overnight, and it reaffirmed the policy of the Communist Party that the success of the Communist movement is dependent on an aggressive science and technology program. On this subject, the Soviet position was poignantly summarized for me by a brilliant young Russian scientist named Victor Zhuravlev, affiliated with the Institute of Space Research of the U.S.S.R. Academy of Sciences, when he said, "We believe that everything can be explained by science and technology!"

One of the principal sources of pertinent intelligence information for the Soviets, unknown to most Americans, is international conferences. Junior KGB intelligence officer Igor Prissevok once told me that the Soviets attend scientific international meetings because it is one of their best means of acquiring information. He explained that receptions, cocktail parties, and similar informal gatherings were valued very highly by the Soviets, more so than the reports and presentations made during formal sessions at these conferences. The caliber of intelligence officers that the Soviets send to international meetings is perhaps the best testimony to their importance. The list includes KGB specialist Igor Milovidov, whose cover was officially blown when he was named by the late GRU Colonel Oleg Penkovskiy; Evgeniy Ivanov, the Soviet intelligence officer involved in the Christine Keeler-John Profumo sex scandal of the early 1960s which overturned the Macmillan government in Britain; KGB Colonel Nikolai Beloussov, the mastermind behind the sex scandal; Georgiy Zhivotovskiy, senior level KGB intelligence officer and assistant to the deputy premier of the Soviet Union; and numerous others. Yuri Riazantsev, a Soviet engineer who is highly regarded by the KGB and works at the Institute of Mechanical Problems of the U.S.S.R. Academy of Sciences, is a KGB watchdog at international meetings. Riazantsev has cultivated friendships with high-level West European scientists during the past several years and on occasion assists senior-level KGB intelligence officer Georgiy

Zhivotovskiy. The results of Riazantsev's work should surface in a few years and there is no reason why the Soviets will not benefit significantly from the propulsion and laser information accessible to Western scientists who have been targeted by the KGB.

On the subject of lasers, it is known that the Soviets are engaged in an aggressive laser death ray weapon development program for the Ministry of Defense, and the KGB has been very active in supporting this effort with its espionage activities in the United States. The KGB's efficiency was demonstrated to me during May 7–10, 1969 when I was in Venice, Italy. During this period my company, the Florida Research and Development Center, was involved in a secret Department of Defense laser program that was given the code name Eighth Card; in a facility of over 6,000 people, only a few engineers knew about the program, and the cost to keep the existence of the program a secret was enormous. On the evening of May 7 I had a drink with "Professor" Georgiy Zhivotovskiy in the lounge of the Ala Hotel. We discussed mostly trivia but he did learn my name and affiliation. Thirteen hours later on May 8 I had dinner with Zhivotovskiy, and he not only knew about my scholastic background and past encounters with Russians, but he knew that the Florida Research and Development Center, a jet and rocket engine company, was up to its ears in lasers. In fact, he pompously invited me to visit the Institute of Nuclear Physics in Novosibirsk so that I "might learn something new about lasers." The only possible explanation was that during the early morning hours of May 8, Zhivotovskiy ran a dossier check on me using the secured facilities of the Russian embassy in Rome and the assistance of his KGB cohorts, Nikolai Beloussov and Vladimir Istomin, who were assigned to the embassy at the time; obviously, there was a laser security leak in the United States and the KGB in Dzerzhinsky Square in Moscow notified Zhivotovskiy about the laser program.

The KGB officers I met while conducting research for this book were noticeably very polished, sometimes arrogant, and looked as if they were either diplomats or movie stars. They were not the cloak and dagger men that one would associate with an Ian Fleming James Bond novel.

Zhivotovskiy was by far the most interesting and influential KGB officer I ever encountered. We met in Yugoslavia in 1967, and we subsequently had dinner and late evening discussions in Venice and Mar del Plata, Argentina, where he was assessing me as a potential spy for the KGB. The Russian intelligence collection effort at international conferences relative to the United States effort is very professional, primarily because they have worked harder and longer in this area than we have. Additionally, all Soviet representatives attending an international conference are accountable to the KGB in one form or another, whereas this is not the case between American attendees and the U.S. intelligence services. Soviet scientists are not only thoroughly briefed on security procedures but they are instructed on the workings of an intelligence organization. Some scientists have been coopted by the KGB and are full-time intelligence officers. The effectiveness of the Soviet intelligence apparatus was again demonstrated to me when I had the unique experience of conversing on separate occasions with a senior-level KGB officer responsible for the Soviet intelligence effort at a particular international meeting, and his counterpart, an American intelligence officer responsible for the American coverage of the same event. When I asked for each man's appraisal of the event, the KGB officer said, "we found it to be very worthwhile"; the American intelligence officer replied, "it was a fiasco." I was able to validate that the American operation was indeed a fiasco, and a waste of taxpayers' dollars, because it was under the control of inexperienced personnel. The ineptness of the American intelligence effort in certain critical areas relative to the Soviets is well-known within private circles of the U.S. intelligence community, and it prompted one civilian intelligence analyst, when he heard that I was working on this book, to write during September 1972 and send me a schedule for international meetings for 1973. He wrote: " ... See the spooks in action—An opportunity to see both (1) amateurs ... Air Force Foreign Technology Division and (2) professionals, U.S.S.R. KGB. Compare their techniques, etc; warning—stay on the sidelines to avoid the cross fire."

It can be stated without qualification that the Soviet intelligence services are more effective in stealing U.S. space, defense, and industrial secrets than the U.S. intelligence services are in stealing Soviet secrets, and the reason for this is that the Soviet closed society with its secret police does a better job of protecting their secrets. It has already been mentioned that Soviet agents can operate practically at will throughout the United States because of our open society, but the key to the Soviet's success is that our most sacred secrets are in the hands of personnel such as engineers, scientists, researchers, and military officers who have not been thoroughly briefed on espionage techniques; they do not comprehend in intelligence terms the significance of being discreet as do their counterparts in Russia. In intelligence terminology, the KGB considers the continental United States to be a picnic, a place where espionage operations can be conducted with relative ease; yet most Americans believe that the Federal Bureau of Investigation is effectively executing its counter-intelligence responsibilities with the ease and efficiency depicted every Sunday evening by Lt. Erskine on the television series, "The FBI." In actuality, the spies who are apprehended by the FBI represent a drop in the bucket compared to the total number of spies performing spy missions in the United States for the Soviet intelligence services. (This number is classified.)

It is worth noting that the Communist Party of the Soviet Union has decreed that its number one espionage target is the United States. The Soviet espionage effort against its number two target, West Germany, is also massive: The Federal Office for the Defense of the Constitution —West Germany's intelligence service—estimated that for the early 1970s there were at least 25,000 Communist spies operating in West Germany, broken down as follows: More than 10,000 Russian agents, 13,000 East German agents, and about 2,000 agents from other Communist countries. It is the objective of these Eastern bloc spies to infiltrate the social, economic, political, scientific, and technical establishments in Germany to obtain valuable unclassified industrial secrets and classified military secrets. The West German Federal Office reported that between 11 percent and 30 percent of the aliens in West Germany are suspected of being involved in espionage or gathering valuable nonclassified information.

Soviet espionage against France and Great Britain have already been clearly documented in the past. In Britain there were the cases of Soviet master spies Guy Burgess, Donald Maclean, Harold Kim Philby, and George Blake; these men penetrated the British intelligence services at the highest levels, spied for the Russians, and eventually escaped to the Soviet Union under bizarre circumstances. Extremely sensitive American secrets passed through their hands, and during his tenure, Philby had personal contact with the late director of the CIA, Allen Dulles. In France, Philippe Thyraud de Vosjoli, the French intelligence officer who assisted the CIA just prior to the Cuban missile crisis, disclosed that the Soviets had penetrated the government of President Charles de Gaulle and the French intelligence service. Public revelations such as these, and statements by the late director of the FBI, J. Edgar Hoover, and officials of the German intelligence services, acknowledge that their organizations are greatly understaffed to counter the massive espionage efforts directed at the West by the Soviets.

The best defense against Soviet espionage is to protect those secrets that are absolutely vital to our national security. But how effective are we in this area? In the following paragraphs I have documented past and current factual conditions in the United States that have led to, or could lead to, a compromise of our national security. Some of the incidents described below are based on firsthand information that I researched for this book, or information provided by concerned Americans. For obvious reasons, in most cases I have withheld the names of the negligent personnel involved.

Scene 1: Scores of civilian intelligence analysts who are assisting the Air Force Foreign Technology Division congregate in California to discuss Soviet weapons systems. Since they are assisting the Foreign Technology Division, the analysts are advised that they could be targets of Soviet espionage agents, and therefore, they must always be alert and take special precautions. Prior to this meeting, however, intelligence officers from the Foreign Technology Division mailed a list of all participants—military and civilian intelligence analysts—to a hotel manager and numerous other persons without taking any precautions whatsoever to protect their identities. If the KGB obtained this list, identifying our best intelligence analysts in the propulsion field, the damage to the nation could be irreparable and numerous covert U.S. intelligence agents would be uncovered as well (i.e., guilt by association with persons on the list).

Scene 2: Revolutionary Airport, a desert airstrip in Jordan, where the commandos of the Popular Front for the Liberation of Palestine (PFLP) hold three hijacked jets and their passengers for ransom. Inside one of the planes is secret U.S. intelligence information that concerns the national defense of the United States. The information should have been transferred by the U.S. Air Force in a military transport aircraft, but for some reason the information was ordered to be delivered via a commercial aircraft. The commandos burn the three planes leaving a pile of rubble. Were the secret documents burned, or are they resting in the secured files of the KGB in Dzerzhinsky Square in Moscow?

Scene 3: The security office of a U.S. aerospace corporation involved in the design and development of secret weapons systems for the Air Force. A security officer responsible for protecting the facility's secrets from foreign agents, and known for his adamant and hard-line toughness on all matters dealing with security, routinely fixes irregularities in Department of Defense security violations for some of the corporation's employees, because exposure of the violation (i.e., leaving classified information unprotected and exposed to personnel who have not been granted security clearances by the Department of Defense) would cast doubt on the competence of the security department and himself. While the damage to the nation could be irreparable if the classified information were ever compromised by foreign agents, this security officer's main concern is to ensure that the facility cannot be legally blamed by the Department of Defense if the security violations are uncovered at a later date.

Scene 4: John F. Kennedy International Airport. A European jetliner discharges 148 passengers including one political refugee who was fortunate enough to have an American sponsor. The refugee, a trained espionage agent, was briefly questioned by the U.S. State Department in a refugee camp, and on the signature of his sponsor was permitted to enter the United States, even though the sponsor was never contacted by any representative of the United States government beforehand.

Scene 5: An American citizen employed in the computer section of an aerospace company, contracted by the government to work on a secret Department of Defense program, collects minute design information about the company's computers and products, and routinely transmits this data to relatives living behind the Iron Curtain by ordinary airmail letters. All of the man's relatives reside behind the Iron Curtain, but he is a bona fide citizen of the United States because his late father was a United States citizen. The man was raised behind the Iron Curtain, but due to a miraculous series of events he "escaped" to the West to resume a new life. The data that the man mails to his "relatives" is vital information that the KGB needs to close the gap in computer technology between the East and West.

Scene 6: An Eastern bloc refugee residing in the United States makes his weekly visit to the Cape Kennedy Space Center-Patrick Air Force Base complex and photographs the personnel and aircraft. He also takes leisurely drives down the Gold Coast of Florida, photographing and taking notes of the Florida Air Defense Network. During his drives across the state he takes "harmless" photographs of the Army Corps of Engineers installations that are off limits to Soviet diplomats. After his pictures are developed by Kodak, the refugee mails them to his "family" so that they can see what America is like.

These photographs are badly needed by the KGB to prepare tactical, strategic, and other intelligence estimates on the United States.

Scene 7: An aerospace facility developing advanced fighter aircraft engines for the Air Force. An aerospace engineer with a secret clearance is permitted to carry a briefcase past two security checkpoints, talk his way past a guard who is supposed to examine the contents of any briefcase taken out of the secured facility, and board a jetliner enroute to an Iron Curtain country. The briefcase was last seen open near a security file filled with several hundred classified defense documents. The security procedures at this facility also permit any employee to use the unmonitored Xerox machines in the evening to copy classified documents. Therefore, copies of secret documents can be removed from the facility at will without any outward sign of an irregularity. This particular facility has received several national awards for excellence in security procedures and Air Force personnel who have visited it have remarked that the facility's security force is infinitely better than U.S. government installations.

Scene 8: The engineering department of an aerospace defense contractor. Project engineers for this aerospace firm were severely reprimanded by the United States Air Force for not keeping abreast of Soviet weapons systems development programs. To comply with the Air Force directive immediately, the project engineers circumvent established Department of Defense security regulations and transmit classified intelligence information about a Russian aircraft by first class mail without appropriate security safeguards. They argue that the security bureaucratic red tape, if followed to the letter, would deter them from complying with the Air Force directive. Voluminous secret government intelligence reports, computer printouts, and design lay-outs, are handled in slipshod fashion and stored in cardboard boxes near a window-sill instead of security-approved combination lock filing cabinets. Compromise of this information by foreign agents would jeopardize the lives of U.S. agents and U.S. pilots who will eventually fly aircraft designed from the information left unguarded in the cardboard boxes.

Scene 9: An aerospace firm has just completed its one-year study of the Soviet space program and defense establishment, based on classified information provided by the Defense Intelligence Agency (DIA) and the Central Intelligence Agency (CIA). Because the study had low priority relative to other programs worked on by the firm, the draft of the intelligence assessment is given to an independent technical publications service company that operates in another town but has been granted a security clearance by the government to prepare such publications. This was done against strenuous objections by the author of the intelligence assessment who maintained that (1) intelligence data must be handled by a minimum number of persons to avoid a compromise in security and (2) the report should be prepared entirely "in-house" by the firm's publication department. A year and a half after the report was published the service company receives a security violation because some of the secret intelligence data and organization charts on the Soviets were found in its files, when all material was supposedly returned to the aerospace firm's security department. There were rumors that a member of the service group visited the Soviet Union before the violation was discovered, but attempts were made by supposedly responsible persons to cover up the incident to salvage the company's security clearance with the government. Compromise of this information to the Soviets would have exposed hundreds of U.S. agents and collaborators and compromised numerous U.S. intelligence operations and classified programs.

Scene 10: A local bar during the weekly Friday evening "happy hour."An aerospace engineer complains about the excessive amount of overtime and weekends that he spends working on a secret Department of Defense laser death ray weapon system. His first-time acquaintance, a writer, tells the engineer about Soviet work in the laser field, especially the possible use of the laser death ray orbital spacecraft and space stations. He asks the engineer if the Soviet work should be taken seriously. The engineer divulges that the work is significant because the United States is thinking of similar applications and the testing of this laser system is currently being conducted at his research center; the work is classified secret, and the power-levels attained by the laser system are also classified, so he cannot discuss it, but the engineer does disclose that his center recently achieved a major breakthrough in cooling technology that permits the power-level of the laser beam to be increased significantly so that it can do more damage to the target without burning out the reflective mirror surface in the laser system. The writer names other engineers and asks about their status. The engineer confirms that the others are still heavily involved in the program, and though he cannot disclose the classified nature of their work, the engineer names other personalities involved in the secret laser program. (This disclosure has been sanitized for presentation here but it is representative of tens of thousands of similar indiscretions that have greatly assisted the KGB and Soviet military planners. In this exchange the engineer (1) could have targeted other U.S. engineers for exploitation by the KGB and GRU and (2) provided information about the status of the laser effort, potential use of the weapon, and information about a breakthrough. The Soviet KGB and GRU routinely assign lower echelon personnel and use Eastern bloc intelligence operatives to frequent lounges, bars, and nightclubs near military bases and aerospace facilities. When ripe targets are found, the KGB takes over the exploitation of the target and assigns a top-notch professional with years of training and experience to the case.)

Scene 11: A conference room in the headquarters building of the Federal Bureau of Investigation. An FBI agent is pleading with his superiors that U.S. counterintelligence efforts are grossly ineffective and more men must be assigned to protect the internal security of the United States from foreign espionage agents. Another official chastizes the State Department for permitting foreigners of unknown backgrounds to enter the United States at will without properly screening the individuals beforehand, thus placing a burden on the internal security forces within the United States, which already have their share of problems. A leading counterintelligence expert is alarmed at the ignorance displayed by FBI personnel on espionage matters; he recommends that FBI field offices upgrade the expertise of their personnel in these areas to be on par with those agents already assigned to the Washington and New York counterintelligence offices. An FBI administrator says they have no jurisdiction over the State Department, and there is nothing that can be done about these grievances because the FBI must operate within its budget.

Scene 12: A briefing room in the main headquarters building of the Central Intelligence Agency. Clandestine CIA agents meticulously present evidence and detailed dossiers of foreign espionage agents who have recently infiltrated the United States, and to their dismay are told that communications and coordination with the FBI have reached an all time low due to an internal rift between the CIA and FBI. Because the CIA is not authorized to conduct counterintelligence operations within the United States (i.e., this is the FBI's chartered responsibility), the internal security of the United States is gravely threatened by known Soviet espionage agents and nothing can be done to improve the situation until petty personnel problems are resolved.

Scene 13: An aerospace engineer with a secret clearance has been routinely removing material and literature from the premises of a renowned research center for years because of a loophole in the center's security procedures: all persons must submit to a security search upon leaving the engineering building unless they elect to take a company automobile shuttle to another of the center's secured building complexes. The engineer, carrying documents in a folder, bypasses the company's security checkpoint in the engineering building by boarding the shuttle. When the engineer reaches his destination, he (1) mingles with his colleagues, who are strolling on the premises during their lunch break, (2) walks to his car and deposits the stolen material, and (3) returns to the engineering building. Numerous employees have known about this loophole, and there is no telling what else has been removed from the facility. (After disclosure of this loophole in the October 16, 1972 issue of the *Palm Beach Post*, the Florida Research and Development Center took internal measures during December 1972 to correct the situation.)

Scene 14: A Defense Intelligence Agency (DIA) installation during the summer of 1972. An Army intelligence officer assigned to test security within the DIA complex photographed DIA personnel during their lunch hour, made his own DIA identity badge, entered the DIA facility, penetrated an unguarded vault containing top-secret documents, requested and obtained classified information, operated the agency's computers, and left the premises with the classified information, unchallenged.

Scene 15: The JANNAF (Joint Army, Navy, NASA, Air Force) Propulsion Meeting in the Jung Hotel in New Orleans during November 27–29, 1972. A noted scientist explains that security measures for the classified sessions at the meeting are so comprehensive that no stone has been left unturned. In fact, the layman could not comprehend how secure the facility is. Meanwhile, a civilian with a secret clearance babbles away in a drunken state in the lounge, while the officials responsible for maintaining security for the meeting compromise information that was forwarded to them in confidence by prominent U.S. scientists and officials. The following information, accessible to anyone, was thrown in a wastepaper basket in the hotel:

A file containing the names of U.S. scientists, their organization, mailing address for classified material, date of birth, place of birth, citizenship, classification level (i.e., whether the man can work with confidential, secret, or top secret documents), type of security clearance granted and by whom, effective date that the security clearance was granted, name of security officer who granted the clearance, classified programs the individual is working on, admission cards which permit one to enter classified propulsion sessions, and other information valuable to KGB operatives, whose business it is to maintain dossiers on U.S. officials with secret clearances. The individuals listed in this file included Ralph Kelley of Aerojet Solid Propulsion Laboratory; Ira Kay of United Aircraft Research Labs; Richard Gompertz of Chrysler Corporation Space Division; John Gregory of NASA-Lewis Research Center; Armond Kaloust of TRW Systems; Paul King of General Electric; Patrick Kelly of McDonnell Douglas Corporation; Ronald Goe of Rocketdyne, a division of North American Rockwell Corporation; Homer Howell, Jr., of the Air Force Space and Missile Systems Organization; and Leon Green, Jr., of the Defense Science Board and the Office of the Director of Defense Research and Engineering, Pentagon. One of the projects listed in this file included the United States Air Force Minuteman ICBM program.

The information on the individuals listed could just as easily have been targeted for the KGB files in Dzerzhinsky Square, where it would be used to assist the KGB in exploiting U.S. scientific and technical personnel. (*See Plate 7.*)

Careless handling of private and sensitive information concerning the U.S. space and defense establishment and the key personalities involved have greatly assisted foreign espionage agents in the performance of their tasks. The compromise of vital information in many cases is never reported because of political reasons (i.e., disclosure would be embarrassing), or because the KGB obtained the information without the knowledge of U.S. security and counterintelligence agents. Scenes 1 through 15 are only a few examples of how our country's most vital secrets and private information are "protected" from the KGB and GRU. The American public has been led to believe that U.S. counterintelligence efforts against the Soviets are adequate, when there is nothing further from the truth. This image can be attributed to a sleek government propaganda campaign, supported by the news coverage of a handful of sensational espionage cases. Americans should be concerned by the tremendous number of competent spies operating on our soil and the lack of appropriate security measures within the United States to protect those secrets that are critical to our national security.

It can be generally concluded that the KGB and GRU continue to steal the innermost space and defense secrets from our government, and they have an in-depth understanding of our strengths and weaknesses. Should the Soviets decide to initiate hostilities against the United States, they will be in the enviable position of knowing exactly what they are doing, and regardless of the outcome of such a conflict, no one could reasonably charge that the KGB and GRU failed them.

SAS-4

To find out what the Soviets do with secrets stolen from the West, we must slip behind the Iron Curtain and take a close look at the Soviet academic-military-industrial-complex that is of major concern to U.S. intelligence analysts.

1972 JANNAF PROPULSION MEETING
(INCLUDING AIAA/SAE ORGANIZED CLASSIFIED SESSIONS)
JUNG HOTEL
NEW ORLEANS, LA

27-29 NOVEMBER 1972

AFAPL/RJA
WRIGHT-PATTERSON AFB, OH

THIS REGISTRATION FORM MAY NOT BE REPRODUCED BUT MAY BE TRANSFERRED WITHIN YOUR ORGANIZATION. ALL APPLICATIONS MUST BE COMPLETED AND RETURNED BY 15 NOVEMBER 1972.*

I. **PERSONAL DATA** (Please print or type)

Name ▇▇▇▇▇ Last, First, Middle

Signature and Title ▇▇▇▇ Captain, USAF

Organization AFAPL/RJA

Mailing Address for CLASSIFIED MATERIAL
AF Aero Propulsion Laboratory
Attn: RJA (Capt ▇▇▇▇
Wright-Patterson AFB OH 45433

REGISTRY No. 00649

Date of Birth 12 May 1940

Place of Birth ▇▇▇▇

U.S. Citizen? Yes
(If NO, give status on reverse of this card)

REGISTRY NO. 00648
JANNAF 1972
PROPULSION MEETING
New Orleans, LA. November 27-29 1972

THIS BADGE REQUIRED FOR ADMITTANCE

THE SECURITY LEVEL OF THE CLASSIFIED SESSIONS IS CONFIDENTIAL. APPROVAL TO ATTEND IS CONTINGENT UPON SATISFACTORY COMPLETION OF THE FOLLOWING CERTIFICATION. SECURITY IS UNDER ARMY CONTROL.

II. **CLEARANCE CERTIFICATION**   □ CONFIDENTIAL

The above named employee of this agency was granted a   XX SECRET   clearance on 1 August 72 (Date)

By 4950/SP, W-PAFB, OH   □ TOP SECRET   and this clearance is still in effect.
(Clearing Agency)

Facility Security Officer
(Signature) Thelma V. Smith
(Printed Name) THELMA V. SMITH
(Title) Staff Office Security Advisor

*Mail To: Chemical Propulsion Information Agency
The Johns Hopkins University/APL
Attn: Mrs. E. M. Daniels
8621 Georgia Avenue
Silver Spring, Maryland 20910

7. FOR USE BY VISITOR CONTROL

11. SECURITY CLEARANCE CERTIFICATION (To be signed by Chief, Visitor Control Section.)

VISITOR ACCREDITATION

PAGE 3 OF 5

6. REMARKS The purpose of this visit is to attend a 1972 Joint Army/Navy/Air Force Propulsion Meeting at New Orleans, Louisiana

10. SPECIAL ACCESS CERTIFICATION (To be signed by DTIC Officer.)

9. NEED-TO-KNOW CERTIFICATION (To be signed by Chief, Originating Office.)

5. PERSON(S) TO BE CONTACTED

8. CONTRACT NUMBER

Chief, Propulsion Division
Directorate of Engineering

NOTE: Instructions for preparation are on the reverse side of the last copy)

VISIT REQUEST

1. CATEGORY   5   OR   6

DATE 9 - NOV 1972

Rec'd CER NOV 1972

PLACE TO BE VISITED:
TO: Chemical Propulsion Information Agency
STREET: John Hopkins University
Applied Physics Lab
CITY AND STATE: 8621 Georgia Avenue
Silver Springs, MD 20910
THRU:

FROM: SAMSO/RPOV
P O Box 92960
Worldway Postal Center
Los Angeles, California 90009

Telephone – 213 – 643-2675

The following named personnel, all U.S. citizens unless otherwise designated in block 6, will visit your facility or installation during the year or on a one-time basis as indicated below. They have a need-to-know for information concerning areas shown in the column titled "Program and Management Area," (Block 4F).

LONG-TERM VISIT THRU _____ (Date)   ONE-TIME VISIT ONLY X   DATE OF FIRST VISIT 27 Nov 1972

GENERAL INFORMATION

OFFICE SYMBOL A	GRADE B	NAME (Last, First, M.I.) C	JOB TITLE	DATE AND PLACE OF BIRTH D	CLEARANCE INFORMATION E	PROGRAM AND MGT AREA
▇▇	LTCOL	▇▇▇▇ Chief, Propulsion Div		28 Jan 1924	460-30-0890 TS 21 Nov 68	Minuteman
▇▇	1LT	▇▇▇▇ Project Officer		21 Nov 1946	378-46-7452 TS 7 Nov 73	Minuteman

WILLIAMS H WLA
CHEMICAL PROPULSION INFORMATION AGENCY 710 863 2561 WRC MSG. NO 085
11-21-72
JOHNS HOPKINS UNIV./APL SILVER SPRING, MARYLAND
ATTN: MRS. E.M. DANIELS

THE FOLLOWING ▇▇▇▇ PERSONNEL WILL BE ATTENDING THE 1972 JANNAF PROPULSION MEETING IN NEW ORLEANS ON NOV 27-28.

NAME-TITLE	DATE & PLACE OF BIRTH	SEC. CLEAR.	AUTH. & DATE
▇▇▇ CHIEF APPL.ENGR	15 FEB 1929 DETROIT, MICH	SECRET	HDQRTS 5TH US ARMY-CHICAGO, ILL 15 MARCH 1963
▇▇▇ CH. DEV. ENGR. AIRCRAFT	19 APRIL 1922 KIRKSVILLE, MO.	SECRET	HDQRTS 5TH US ARMY-26 JULY 1962
▇▇▇ CH. ANALY. ENGR	1 JAN 1927 NORTH YORK, ONT. CANADA CANADIAN CITIZEN A-11-104-169	SECRET	HDQRTS 5TH US ARMY-17 OCT 1962

▇▇▇ NAVY SECRET FACILITY CLEARANCE GRANTED 11 FEB 1960 BY HDQRTS 5TH US ARMY, CHICAGO, ILLINOIS. WRITTEN VERIFICATION WILL FOLLOW.

PLATE 7   Compromise? "Santized" Presentation of Private Records Thrown in Wastepaper Basket during Classified Propulsion Conference in New Orleans (November 1972)

Chapter 3

The Soviet Science and Technology Establishment

United States secrets are delivered to KGB and GRU personnel in Moscow by various methods. One of the simplest is by hand-carried diplomatic pouch from the Soviet embassy in Washington; diplomatic mail and documents are immuned from examination by U.S. law enforcement agencies, just as diplomats are immuned from arrest (diplomatic immunity). In Moscow, the secrets are delivered to the KGB headquarters in Dzerzhinsky Square, the GRU in the building of the Ministry of Defense on the Moscow River, other KGB and GRU centers, and a monstrous building that has been under construction since the late 1960s—the Institute of Space Research of the U.S.S.R. Academy of Sciences. A KGB science and technology specialist once described the Institute of Space Research to me as "a skyscraper lying on its side"; it is about half a kilometer long (about a quarter of a mile) and fourteen stories high. The institute will direct all Soviet space activity from the first seven floors (this is discussed in later sections) and will house support personnel and facilities in the top seven floors for evaluating foreign, and primarily United States, science and technology. This portion of the Institute of Space Research will represent the most advanced technical intelligence complex in the world for exploiting the space and defense secrets of other nations. It includes a massive secret library containing raw technical intelligence reports compiled by KGB and GRU operatives and their agents throughout the world, an enormous library of the significant unclassified technical publications of other nations, a massive translation department headed by multilingual personnel and computers that translate American publications into Russian, and teams of intelligence analysts who evaluate both publications and raw intelligence information and prepare finished intelligence reports for the KGB, GRU, Ministry of Defense, Academy of Sciences, and State Committee for Science and Technology—organizations that are discussed in later paragraphs. One member of this technical library system is Youri Zonov, the chief of the Department of Scientific Information at the Institute of Space Research. Zonov, a friendly dark-haired man in his thirties who could pass as a New York engineer, was originally assigned to the United Nations where he helped draft the Space Treaty. He has assisted senior KGB intelligence officer Georgiy M. Zhivotovskiy, who heads Soviet intelligence collection operations at various international conferences.

In essence, the Soviets have done something that numerous frustrated U.S. intelligence analysts have been unsuccessful in getting the U.S. government to do: created an efficient complex where significant intelligence information concerning foreign technological developments can be immediately incorporated into the design of one's own weapons systems. The Soviet solution to the problem was to house their technical and scientific intelligence departments in the same building that houses the executives of the Soviet space program and other technological developments (i.e., the Institute of Space Research). In the United States, the intelligence community is so rigidly compartmentalized and separated from the science and technology community, both physically and by procedures, that technical intelligence information concerning the Soviets is not being transmitted in time to the users of the intelligence (i.e., scientists and engineers in the aerospace industry and research establishments who develop comparable systems). In many instances the intelligence data on the Soviets is obsolete by the time the designers of U.S. weapons systems get to see it.

Soviet intelligence assessments of American weapons systems are generally more reliable than American assessments of Soviet weapons systems. The reason for this is that the Soviets not only have more information to work with, because of our free press and their competent espionage operations, but their intelligence community has a very close working relationship with scientists and engineers who are up-to-date with the latest technological developments (i.e., the state-of-the-art). Since it is their business to design, develop, test, and evaluate weapons systems, Soviet scientists and engineers are the most qualified to evaluate comparable work done by their U.S. counterparts.

In the United States intelligence analysis is done mainly by the intelligence community. In the areas of science and technology, it is known in private circles at the highest levels that the available U.S. intelligence analysts, such as those employed by the Air Force Foreign Technology Division, are not qualified to do the work because they lack experience and are not familiar with the state-of-the-art. Yet these intelligence analysts, qualified or not, are responsible for determining the design specifications of billions of dollars of weapons systems developed in the United States. It is even more alarming that the CIA, whose budget is dwarfed by the military intelligence services, is in fact leaving the bulk of scientific and technical intelligence work to the military intelligence services, whose competence is also questioned by leaders in the aerospace industry, who are rightfully more qualified to evaluate the Russian work.

In summary, the Soviets have created a complex whereby significant technical intelligence concerning the West is diverted immediately to the users of the intelligence—the Soviet political-espionage-academic-military-industrial complex. This complex consists of a politburo, made up of men who support both an aggressive espionage effort abroad and a science and technology program at home; an espionage establishment made up of competent KGB and GRU officers; and a scientific and technical establishment located mainly in the Moscow area.
SAS-5

The planning and coordination of Soviet science and technology, and the integration of the fruits of this work into the Soviet space and defense establishment are the responsibility of the Soviet government —the Council of Ministers of the Soviet Union, headed by Premier Aleksei Kosygin. There are scores of ministries and state committees that report to the Council of Ministers, and the head of each ministry and state committee is a member of the council. Two significant organizations responsible to the Council of Ministers are (1) the State Committee for Science and Technology and (2) the Academy of Sciences of the Soviet Union. During President Nixon's Moscow summit of May 1972, Americans had the opportunity to see the heads of these branches of the Soviet government: president of the Academy of Sciences Mstislav V. Keldysh, the senior white-haired scientist who has been in that position since 1961, met President Nixon at Vnukovo Airport and was the Soviet promoter of the joint spaceflight mission between the two countries that will involve the link-up of an Apollo spacecraft with a Soyuz spacecraft in 1975; chairman of the State Committee for Science and Technology Vladimir A. Kirillin, who was vice president of the Academy of Sciences during 1963–1964 before becoming chairman of the State Committee in 1965, was the Soviet political official who cosigned the Science and Technology Agreement with Secretary of State William P. Rogers in the Kremlin's St. Vladimir Hall during the Moscow summit.

The academy is chartered to develop the sciences in the Soviet Union; the work of its institutes and design bureaus is mostly basic research-oriented, which means that ideas are tested in laboratories. The State Committee for Science and Technology, which is made up of scientists from the academy, scientists and engineers from Soviet industrial establishments, and intelligence officers, is responsible for developing the technology within the Soviet Union, and the work of its institutes and design bureaus is mostly applied research-oriented. The ideas successfully researched by the academy are studied in more depth, and the State Committee for Science and Technology seeks engineering solutions to technical design and hardware problems rather than scientific solutions to ideas. It is shown in later paragraphs that the Academy of Sciences is in fact responsible for the Soviet space program, and the State Committee for Science and Technology has more administrative authority than the academy and is responsible for coordinating the efforts of numerous ministries that support the Soviet defense establishment, including some of the work performed by the academy.

Though the academy and State Committee for Science and Technology report to Kosygin's Council of Ministers, they are closely controlled by the Communist Party, and this is possible because of the closely-knit political-academic complex in the Soviet Union. For example, not many Americans are aware that General Secretary Leonid Brezhnev is a metallurgical engineer and Premier Aleksei Kosygin is a textile engineer, and that approximately half of the Central Committee of the Communist Party has a background in science and technology. President of the Academy of Sciences M. V. Keldysh and chairman of the State Committee V. A. Kirillin are members of the Central Committee. In the Academy of Sciences, two of Keldysh's top assistants, M. Lavrentyev and Vadim A. Trapeznikov —Trapeznikov is also Kirillin's deputy on the State Committee —belong to the Central Committee, and it is known that these men exert considerable influence with Leonid Brezhnev, Aleksei Kosygin, and other members of the ruling Politburo, especially when the academy's budget is reviewed. The importance of the State Committee for Science and Technology to the Soviet government was demonstrated when the Central Committee and Council of Ministers jointly decreed in 1961 that the chairman of the State Committee for Science and Technology (called the State Committee for the Coordination of Scientific Research in those days) would also hold the title of deputy premier of the Soviet Union. In other words, Vladimir Kirillin is Aleksei Kosygin's deputy on the Council of Ministers in addition to being chairman of the State Committee for Science and Technology (which reports to Kosygin's Council of Ministers.) To further intertwine the relationships, Aleksei Kosygin's son-on-law, Dzherman M. Gvishiani, is a vice chairman of the State Committee (i.e., Kirillin is Kosygin's deputy; Kosygin's son-in-law is Kirillin's deputy).

My last personal contact with Georgiy Zhivotovskiy was in Argentina in 1969, when he was both a member of the presidium of the Academy of Sciences (discussed in later paragraphs) and a senior-level KGB intelligence officer. In 1970, my contacts at the Russian embassy in Rome and the Central Aerohydrodynamics Institute in the Moscow area (discussed in later paragraphs) notified me that Zhivotovskiy was promoted to Vladimir Kirillin's State Committee for Science and Technology, and Zhivotovskiy is currently Kirillin's chief assistant, which places him above Kosygin's son-in-law, Gvishiani. Zhivotovskiy's promotion also brings out the KGB involvement in the Soviet academic establishment. Gvishiani (Kosygin's son-in-law), who also holds the title of director of the Foreign Relations Department within the State Committee for Science and Technology, has very close ties with both the KGB and GRU. This has been documented in publications such as *The Penkovskiy Papers* (1965), by the late Colonel Oleg Penkovskiy, and *Contact on Gorky Street* (1968), by British agent Greville Wynne, who not only delt with Gvishiani personally but was apprehended by the KGB because of his involvement with Penkovskiy. Wynne was subsequently charged with espionage and sentenced to the Lubyanka Prison in the KGB building in Moscow. He was returned to Britain in exchange for Soviet spy Gordon Lonsdale in 1964.

> In summary, a handful of men in the Politburo have established the machinery to control Soviet science and technology by assigning trusted and capable scientists to key administrative positions, and rewarding them with membership in the Central Committee of the Communist Party. Members of the Politburo maintain personal contact with the leaders of the Soviet academic establishment, and the involvement of the KGB and GRU is noticeable as well. (See Plate 8.)
> **SAS-6**

Americans have heard of the Academy of Sciences of the Soviet Union, but the extent of most people's knowledge of this significant body ends there. In the paragraphs below I have briefly summarized the organization of the academy and identified those scientists who manage the Soviet space and defense programs that would be of interest to the reader. The Academy of Sciences is made up of a General Assembly of approximately 600 scientists, frequently called "academicians." About 200 members of the General Assembly are full or voting members; they are directors of institutes and design bureaus (research and development facilities) of the academy, and some academicians manage certain aspects of the Soviet space program. Full members belong to the academy for life. The remaining members of the General Assembly are candidate or non-voting members; they are also prominent members of the Soviet scientific establishment. Some are directors of institutes but they do not participate in crucial matters brought before the academy that require a vote. Depending on their specialty, academicians are assigned to one of five major sections that are broken down further into fifteen departments of the Academy. The titles of these departments include the following: Mathematics, General and Applied Physics, Nuclear Physics, Earth Sciences, Economics, and Languages and Literature. Each of the five major sections are headed by a vice president of the academy, who reports directly to M. V. Keldysh, the president. There are two principal bodies within the Academy of Sciences that are worth mentioning. The first is the presidium, which directs the overall effort of the academy and advises the Soviet government. The presidium includes the heads of the fifteen departments of the academy and high-level officials of the KGB, such as Georgiy Zhivotovskiy, before he was transferred to the State Committee for Science and Technology. When I talked with

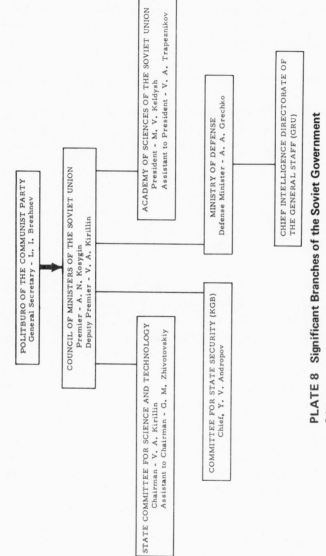

POLITBURO OF THE COMMUNIST PARTY
General Secretary - L. I. Brezhnev

COUNCIL OF MINISTERS OF THE SOVIET UNION
Premier - A. N. Kosygin
Deputy Premier - V. A. Kirillin

STATE COMMITTEE FOR SCIENCE AND TECHNOLOGY
Chairman - V. A. Kirillin
Assistant to Chairman - G. M. Zhivotovskiy

COMMITTEE FOR STATE SECURITY (KGB)
Chief, Y. V. Andropov

ACADEMY OF SCIENCES OF THE SOVIET UNION
President - M. V. Keldysh
Assistant to President - V. A. Trapeznikov

MINISTRY OF DEFENSE
Defense Minister - A. A. Grechko

CHIEF INTELLIGENCE DIRECTORATE OF
THE GENERAL STAFF (GRU)

PLATE 8 Significant Branches of the Soviet Government
SAS-7

Zhivotovskiy on the terrace of a hotel on the Grand Canal in Venice one bright morning, he made the following interesting revelation: The Soviet government has never refused to release reserve funds (i.e., funds in excess of the academy's budget) when requested to do so by the presidium of the academy. According to Zhivotovskiy, the presidium, which meets every Friday, controls the budget allocations within the academy and "is willing to gamble." In what I thought was a significant point, relative to similar scientific endeavors in the United States, Zhivotovskiy said, "If you pursue only that which you understand, soon you will have no science ... You must be willing to gamble in science." He noted that the presidium encourages dual effort and different methods of approach to achieve a single objective.

The other significant body within the academy is called the Commission for the Exploration and Use of Outer Space. The chairman of this commission is aging Academician Anatoly A. Blagonravov. The commission consists of mostly space-oriented academicians, KGB intelligence officers (Zhivotovskiy was also a member of this commission; KGB specialist G. S. Balayan is currently the scientific secretary of the commission), and space and defense specialists from state committees and ministries that also report to the Council of Ministers (other state committees and ministries are mentioned in later paragraphs). The Commission for the Exploration and Use of Outer Space advises and guides the overall direction of the Soviet space program, but management of the space effort and specific projects is left entirely to subordinate councils and managers from the academy. Zhivotovskiy described the commission's role in the following manner: "There are many (space) projects under the guidance of the commission, but each project has one manager."

The Institute of Space Research of the Academy of Sciences, mentioned earlier, is headed by tall and white-haired Georgiy Petrov, perhaps one of the most respected scientists in the Soviet Union. Petrov, who was elected to the position by the full members of the academy and has been greatly responsible for the Soviet space successes to date, told me that the presidium of the academy created the Institute of Space Research because the academy's institutes were much too involved with their own problems and were not adequately coordinating their activities with the Soviet national space effort. The catalyst for this decision was the untimely death in 1966 of Academician Sergei P. Korolev, the brains behind the Soviet space program from its inception. With his exceptional technical and managerial talents, Korolev coordinated the work of hundreds of independent scientific and technical collectives within the Soviet Union. In essence, Korolev was Russia's answer to the National Aeronautics and Space Administration (NASA), which manages our space program. The Institute of Space Research is now the Soviet version of NASA and is responsible for both the Soviet manned and unmanned space programs discussed in later sections. One of three deputy directors of the institute, who was selected to this position by the presidium of the academy, is Academician Jouli Khodarev, a portly scientist with a good sense of humor and a bear-like appetite. With a KGB intelligence officer monitoring our conversation, Khodarev told me that the presidium also selected the department heads of the institute, and these men were free to select their own personnel, but Director Petrov had veto power over all selections. One of the principal objectives of the Institute of Space Research is to plan space experiments for Soviet orbital laboratories that will be launched in the coming decades.

Plate 9 provides a general summary of the organization of the Academy of Sciences of the Soviet Union, some of its key administrative bodies, and its chain of command with the Communist Party. The noteworthy points are: (1) the Academy of Sciences is responsible to Kosygin's Council of Ministers, which executes the policies of the Communist Party; (2) high-level administrators of the academy such as Keldysh, Trapeznikov, and Lavrentyev (not shown) are members of the Central Committee of the Communist Party; (3) the presidium of the academy controls the budget allocations within the academy and advises the Soviet government; (4) the Commission for the Exploration and Use of Outer Space is responsible for the general direction of the Soviet space effort per orders from the Communist Party, but leaves the administrative and technical details to the Institute of Space Research, separate space councils, and managers; and (5) the Institute of Space Research coordinates the activities of the academy's institutes, is responsible for the Soviet manned and unmanned space programs, and can be thought of as the NASA of Russia.
SAS-8

The defense of the Soviet Union is the responsibility of the Ministry of Defense, headed by Defense Minister Marshal Andrei A. Grechko,

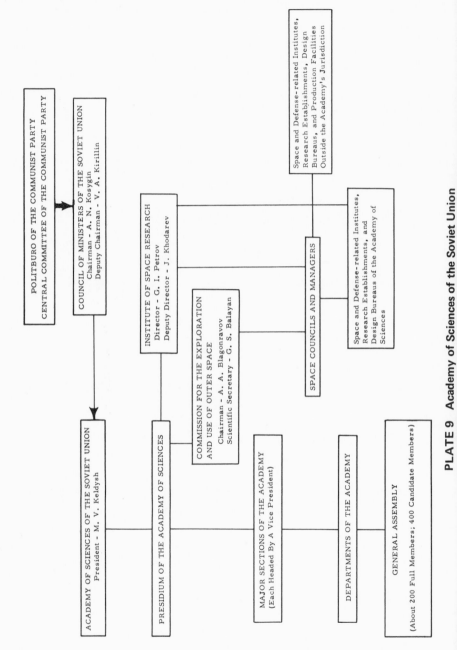

PLATE 9 Academy of Sciences of the Soviet Union

POLITBURO OF THE COMMUNIST PARTY
CENTRAL COMMITTEE OF THE COMMUNIST PARTY

COUNCIL OF MINISTERS OF THE SOVIET UNION
Chairman - A. N. Kosygin
Deputy Chairman - V. A. Kirillin

ACADEMY OF SCIENCES OF THE SOVIET UNION
President - M. V. Keldysh

PRESIDIUM OF THE ACADEMY OF SCIENCES

INSTITUTE OF SPACE RESEARCH
Director - G. I. Petrov
Deputy Director - J. Khodarev

COMMISSION FOR THE EXPLORATION
AND USE OF OUTER SPACE
Chairman - A. A. Blagonravov
Scientific Secretary - G. S. Balayan

MAJOR SECTIONS OF THE ACADEMY
(Each Headed By A Vice President)

DEPARTMENTS OF THE ACADEMY

GENERAL ASSEMBLY

(About 200 Full Members; 400 Candidate Members)

SPACE COUNCILS AND MANAGERS

Space and Defense-related Institutes,
Research Establishments, and
Design Bureaus of the Academy of
Sciences

Space and Defense-related Institutes,
Research Establishments, Design
Bureaus, and Production Facilities
Outside the Academy's Jurisdiction

64

who is also a full member of the ruling **Politburo**. The Ministry of Defense reports to Premier Aleksei Kosygin's Council of Ministers, but it is no secret that the Politburo of the Communist Party holds the real power, and many of Grechko's deputies belong to the Central Committee of the Communist Party. Marshal Grechko's counterpart in the United States is the secretary of defense. The Ministry of Defense consists of five principal and independent departments: (1) Air Force, (2) Navy, (3) Ground Forces, (4) Air Defense Forces, and (5) Strategic Rocket Forces. An interesting feature of the Soviet defense structure is that the Strategic Rocket Forces operate the Soviet intercontinental ballistic missile (ICBM) force, and this department is directly accountable to Minister of Defense Grechko, not the Department of the Soviet Air Force. For comparison, the United States Strategic Air Command is responsible for the U.S. ICBM force, but it is one of many commands within the Department of the Air Force. In essence, the Soviets regard their long-range ballistic missile weapons systems significant enough to warrant the creation of the Strategic Rocket Forces; this provides the Communist Party more visibility over its ICBM program since the Strategic Rocket Forces are not lost within the bureaucracy of their Air Force.

Until his death in February 1972, Commander-in-Chief of the Strategic Rocket Forces Marshal Nikolai I. Krylov was one of the most vociferous spokesmen for the Ministry of Defense. Krylov, who was instrumental in the development of the Strategic Rocket Forces, repeatedly claimed that the United States was developing a first strike ICBM capability. The Strategic Rocket Forces are also responsible for launching manned and unmanned space vehicles for the Academy of Sciences, which manages the Russian space program. Thus, the military plays a predominant role in the Soviet space effort, and as is shown in later sections, it will continue to do so. For comparison, the U.S. unmanned space launches to the moon and planets, and the manned Mercury, Gemini, and Apollo programs were managed by the National Aeronautics and Space Administration, our civilian space agency.

Two points remain to be said about the Ministry of Defense. First, the GRU, the military branch of the Soviet intelligence services, reports to the General Staff of the Ministry of Defense. As was mentioned earlier, GRU intelligence officers are watched closely by their "neighbors," the KGB, to give the Politburo visibility and control over the clandestine operations of the Ministry of Defense. The second point is that Dmitry F. Ustinov, the candidate member of the ruling Politburo of the Communist Party (refer to Table I), who is sometimes called "the heavy industry czar," has close working relationships with the ministers in the Soviet Union who produce space and defense weapons, and this list includes the heads of the Academy of Sciences (Keldysh) and State Committee for Science and Technology (Kirillin). Plate 10 provides a general summary of the organization of the Ministry of Defense, and shows that the ministry is responsible to both the Politburo of the Communist Party and the Council of Ministers.

To place the organization of the Soviet academic-military-industrial complex in its proper perspective, it should be reiterated that the State Committee for Science and Technology, Academy of Sciences of the U.S.S.R., and Ministry of Defense, represent only a few of many state committees and ministries that report to Kosygin's Council of Ministers. But they are among the most important. Plate 11 shows the general administrative organization of this complex, and in most cases the titles of the ministries and state committees listed describe their basic function(s).

The State Planning Committee, commonly referred to as GOSPLAN by the Soviets, is responsible for the overall annual and five-year economic plan of the Soviet Union; it allocates the budgets of the other ministries and state committees shown in Plate 11, but Kirillin's State

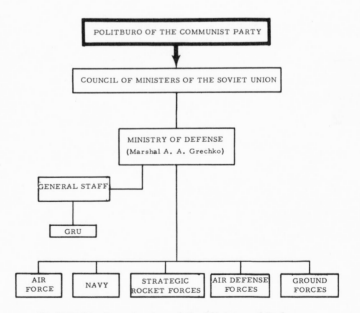

PLATE 10   Organization of the Ministry of Defense
SAS-10

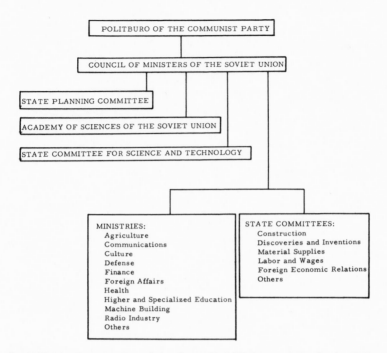

PLATE 11 Organization of the Council of Ministers of the Soviet Union
SAS-11

Committee for Science and Technology has final say over matters that concern the Soviet space and defense complex.

The Ministry of Aircraft Production, which has been headed for a quarter of a century by P. V. Dimentiev, who is a member of Kosygin's Council of Ministers, coordinates its effort with the Ministry of Defense and is responsible for the development and production of Soviet civilian and military aircraft. The institutes and design bureaus under this ministry are among the finest in the world and have recently developed advanced military aircraft that are of major concern to U.S. intelligence specialists. The Ministries of Machine Building and General Machine Building are very much involved in the development of Soviet missile and space booster systems.

It should be noted that relative to the General Secretary of the Communist Party (Brezhnev) and Chairman of the U.S.S.R. Council of Ministers (Kosygin), the President and Vice President of the United States have much less political and executive control over matters that involve the U.S. space and defense establishment; the Congress and public literally determine the course of our nation's destiny. This point was succinctly made clear to me in a cafe in the Old Fortress of Dubrovnik by a serious-minded KGB intelligence officer who enjoyed lauding the achievements of the Soviet Union. The officer said that the Soviet defense complex was better organized than that in the United States because the Soviets planned better. He noted that the Soviets would select individuals with exceptional capabilities and give them dictatorial control over major programs because they believe that a task should be managed by a few competent men rather than a large number of mediocre men. According to the officer, because Brezhnev and Kosygin have the authority to formulate and execute policy, the space and defense competition between the two superpowers is between the Soviet collective leadership and the people of the United States, who are not sufficiently educated on national defense matters to intelligently plan a congruous low-cost program that would hold up over several decades. He noted, for example, that in the United States, people elect officials who appoint managers to look after the aerospace establishment. But after several years, Americans become frustrated with the officials that they elected, so they elect others. The new officials and politicians appoint still other managers. Thus, the continuity of U.S. defense programs is destroyed and the new managers require time to get acquainted with their jobs. This sometimes causes delays, redirection of effort, and an increase in operating cost, with no end in sight. On the other hand, the Soviet academic-military-industrial complex depends on (1) the delegation of dictatorial powers to a few competent managers and (2) continuity of operations by minimizing the turnover in managerial personnel.

In summary, in addition to his foreign policy obligations as the Premier of the Soviet Union, Aleksei Kosygin is the Chairman of the Council of Ministers of the Soviet Union which is responsible for the economy, science, technology, and defense of the Soviet Union. These tasks are executed by numerous state committees and ministries that report to Kosygin, but the Communist Party maintains visibility over all programs and is assured that its directives are being carried out, since leading members of the Central Committee hold key administrative positions throughout this complex and provisions have been made for minimizing the turnover of competent personnel.

SAS-12

Chapter 4

The Soviet Military-Industrial Complex

The deployment of a weapon system is preceded by extensive research, design, development, test, evaluation, and production of the system. Depending on the complexity and importance of the weapon, it sometimes takes between three and ten years for it to reach operational hardware status from the time the weapon was conceived on paper. In other words, the weapons systems that are currently operational in the United States and the Soviet Union are the product of the scientific and technological base that existed in the facilities and institutes in each country during the 1960s; conversely, the space and weapons systems that will be operational in the 1980s and 1990s will be the product of the scientific and technological base available in each country today. With the complexity of weapons systems increasing each year, the time required to research, design, develop, test, evaluate, and produce the system will also increase. A significant point that has not been made clear to Americans is that when a superpower finds itself in an inferior military or strategic position because its enemy has deployed technologically superior weapons, the situation cannot be altered overnight. While the solution to closing a technology gap is the investment of more of a nation's resources in science and technology over several years, this is not always possible. The most feared advances in weaponry that result from a technology gap is a technological breakthrough. Hiroshima and Nagasaki bear the remnants of the United States' technological breakthrough in the 1940s —the atomic bomb. The Japanese would have needed years to develop a similar weapon, but the war ended within days after the bombings, and the rest is history.

One of the reasons for the existence of the Central Intelligence Agency (CIA) and the Defense Intelligence Agency (DIA) is to advise the President and the Department of Defense of Soviet activities that could lead to a technological Pearl Harbor. To perform this task, one must have knowledge of the facilities that play important roles in the Soviet space and defense establishment. The paragraphs below identify the personnel and facilities of the Soviet military-industrial complex that are closely guarded by the KGB and belong to the previously mentioned state committees and ministries of Premier Aleksei Kosygin's Council of Ministers.

Weapons systems development programs in the Soviet Union are the primary responsibility of government-owned design and test facilities called experimental design bureaus. Whereas this type of work in the United States is performed by aerospace corporations that are headed by administrators who promote the development of greatly diversified products to accommodate military and civilian markets and the whims of the controlling interests of the corporation, Soviet experimental design bureaus are headed by dictatorial chief designers who work only in the specialized areas defined by the Communist Party. It is common for U.S. engineers to work for several aerospace corporations during their working lives due to changing conditions in the job market (i.e., contract awards, layoffs, personal preferences, etc.), but in Russia engineers generally work in the same design bureau throughout their working lives; the Soviet government rigidly manages aerospace programs by controlling the job assignments of its engineers. The most significant design bureaus in the Soviet Union are located in or near Moscow and each is named after a living (or dead) chief designer who has headed the bureau. Table III lists some of the better known design bureaus in the Soviet Union; the specialty of each bureau is shown in parentheses next to the chief designer's name. Most of the Russians listed are, or were, highly respected leaders in their field. For example, Sergei P. Korolev, the mastermind of the Soviet space program, headed the bureau that designed and developed the major ICBMs and space launch vehicles in the current Soviet inventory. This included the rocketship that launched Cosmonaut Yuri Gagarin, the first man in space. When Korolev died of stomach cancer in 1966 the leadership of his bureau was reportedly distributed among other Soviet chief designers, including M. K. Yangel, listed in Table III.

The current chief of the Soviet liquid propellant rocket propulsion effort, the rocket engines used in Soviet ICBMs and advanced launch vehicles, is Academician Valentin P. Glushko, who can be considered to be the Wernher von Braun of the Soviet Union; his official title is chief designer of rocket engines of the Soviet Union. Glushko's design bureau is located in Khimki, near Moscow, and is guarded by the most comprehensive security measures imaginable. The closest counterpart to Glushko's design bureau in the United States is Rocketdyne, a division of North American Rockwell Corporation, located in Canoga Park, California. Rocketdyne is currently contracted by NASA to

PRINCIPAL SOVIET EXPERIMENTAL DESIGN BUREAUS

MISSILES AND SPACE SYSTEMS DESIGN BUREAUS

| KOROLEV | (ICBMs and Space Launch Vehicles) |
| YANGEL | (ICBMs and Space Launch Vehicles) |

ROCKET ENGINE DESIGN BUREAUS

GLUSHKO	(Liquid Propellant Rocket Engines for ICBMs and Space Launch Vehicles)
DUSHKIN	(Rocket Engines for Tactical Missiles)
ISAYEV	(Liquid Propellant Rocket Engines for Spacecraft)

AIRCRAFT DESIGN BUREAUS

ANTONOV	(Transport Aircraft)
BERIEV	(Transport Aircraft)
ILYUSHIN	(Transport Aircraft)
KAMOV	(Helicopters)
MIKOYAN	(Fighter Aircraft)
MIL	(Helicopters)
MYASISHCHEV	(Bombers)
SUKHOI	(Fighter Aircraft)
TUPOLEV	(Transport Aircraft, Bombers, Fighter Aircraft)
YAKOVLEV	(Transport Aircraft, Bombers, Fighter Aircraft)

AIRCRAFT ENGINE DESIGN BUREAUS

BONDARYUK	(Ramjet-Scramjet)	(Note: These are technical names for the
ISOTOV	(Turboshaft)	type of aircraft engines designed by the
IVCHENKO	(Turbofan)	indicated bureaus. Ramjet and Scramjet
KLIMOV	(Turbojet)	engines will be used in advanced aircraft
KUZNETSOV	(Turbofan)	in the late 1970's and 1980's; turboshaft
MIKULIN	(Turbojet)	engines are used in helicopters; turbofan
LYULKA	(Turbojet)	engines are used in transports, bombers,
SHVETSOV	(Piston)	and some fighter aircraft; turbojet engines
SOLOVIEV	(Turbofan)	are used in fighter aircraft; piston engines
TUMANSKIY	(Turbojet)	are used in most propeller aircraft.)

TABLE III Principal Soviet Experimental Design Bureaus
SAS-13

develop the rocket engines for the U.S. space shuttle reusable launch vehicle, which should become operational in 1980.

In June 1971 the Soviets announced the death of Dr. Aleksei M. Isayev, the head of the design bureau that developed the rocket engines for Soviet orbital spacecraft systems. Though Isayev was chief of his own design bureau, he reported to Glushko.

Because of the closed nature of the Soviet rocket industry, the experimental design bureaus affiliated with it are not as well-known as those affiliated with their aircraft industry: Soviet rockets, for example, are launched and recovered in the Soviet Union, and those seen in Moscow parades by news reporters, scientific attaches, and Western agents are frequently shrouded to conceal their design characteristics. Soviet transport and bomber aircraft, however, are routinely observed in the West and flying over international waters. Their jet fighters are based in combat zones that subject them to possible surveillance, capture, or destruction, as was the case in the 1967 Israeli Six-Day War, when Western intelligence specialists acquired a wealth of design information about Soviet aircraft and engine systems. For these and other reasons, more is known about Soviet aircraft design bureaus and their design techniques. Some of the better known aircraft design bureaus are mentioned below, because it is shown in later sections that they could plan an active role in the Soviet space program in the future.

The Mikoyan Design Bureau is noted for its MiG series of fighter aircraft. During the Korean War American pilots had great success in their encounters with the MiG 15 jet fighter; in air-to-air combat about 12 Russian-made fighters were destroyed for every American fighter shot down. In the air war over North Vietnam, the 12 to 1 air superiority ratio in favor of the United States was reduced to about 2.5 to 1. In other words, five Russian planes were shot down in air-to-air combat for every two American planes. In Vietnam, American pilots were confronted with the Mikoyan MiG-21, a small hot rod jet fighter that is much better than the record shows because of the inferiority of the North Vietnamese pilots. The Mikoyan Design Bureau has also built the MiG-25 (note: formerly designated MiG-23), a jet fighter that carries the NATO code name FOXBAT, which is considered by U.S. experts to be the most advanced operational fighter in the world for its time period; it was flight-tested in 1965 and deployed recently in the Middle East. The Mikoyan Design Bureau's closest counterpart in the United States is the McDonnell Douglas Corporation in St. Louis, Missouri, which was contracted by the Air Force in December 1970 to develop the F-15 air superiority fighter, America's first major fighter aircraft development program in about two decades.

The Tupolev Design Bureau is accredited with the design of the world's first passenger supersonic transport (SST), which has been designated the TU-144. The Soviet SST can travel twice the speed of sound or roughly 1,500 miles per hour. The Boeing Company in Seattle, Washington, which designed the U.S. SST (cancelled by Congress in 1971) is the Tupolev Design Bureau's counterpart in the United States for passenger jet aircraft.

In September 1971 the Department of Defense announced that the Tupolev Design Bureau was developing a long-range supersonic bomber weapon system, NATO code name BACKFIRE, that can travel about 1,500 miles per hour and has a range which permits it to reach vital strategic targets in the United States. Tupolev's counterpart for this effort in the United States is the department within North American Rockwell Corporation that is designing the B-1 long-range supersonic bomber weapon system, which lags the Soviet effort by about five years. In December 1972 the Tass News Agency reported that 84-year-old Andrei Tupolev died, and is survived by his aircraft designer son, Aleksei Tupolev.

The work of the Kuznetsov and Tumanskiy Aircraft Engine Design Bureaus is very highly regarded by Western experts. The Kuznetsov Design Bureau developed the engines for the Soviet SST, and it has been confirmed by reliable sources that Kuznetsov modified the SST engines, developed for passenger service, for use in the aforementioned Tupolev long-range supersonic bomber (BACKFIRE). General Electric's Aircraft Engine Group in Cincinnati, Ohio is developing the engines for the North American Rockwell B-1 supersonic bomber, and their jet engine development effort parallels the work conducted by the Kuznetsov Design Bureau.

The jet engines designed by Tumanskiy power some of Russia's most advanced fighter aircraft. The innovations of the Tumanskiy Bureau are revered by prominent U.S. propulsion engineers, who often attempt to streamline U.S. approaches to engine design by incorporating Tumanskiy's design philosophy in their work. As an example, during a typical debate between engine designers from competing U.S. aerospace corporations and an Air Force adviser, one engineer threw up his arms in frustration and exclaimed, "If we make that change, we'll still be three years behind Tumanskiy!"—a comment that depicts the respect U.S. engine designers have for this particular Russian. Pratt &

Whitney Aircraft, a division of United Aircraft Corporation, is developing the engines for the McDonnell Douglas F-15 air superiority fighter and is the Tumanskiy Design Bureau's counterpart in the United States.

It would be appropriate at this point to compare operating principles between U.S. corporations and Russian design bureaus. In the United States it is common for two or three aerospace corporations to vie for a government contract. When the contract is awarded to one of the competitors, the losing corporations are often forced to lay off personnel. This usually breaks up experienced design teams because of budgetary considerations rather than by choice. In the Soviet Union, as one Russian engineer explained it to me, the profit motive is removed from the design competition. Two or three bureaus vie for a contract; the contract is awarded to the bureau with the best design; the losing design bureaus are then given money to work on other technical programs that the Communist Party believes would benefit the Soviet Union. The net result is to reward the best design bureau with a major contract (this is a most prestigious event in the Soviet Union, because scientists, doctors, and engineers are revered and afforded special privileges, despite talk about a classless society), while the losing design teams are kept intact for the next competition. The Soviets have found that design experience and a low turnover in personnel are major factors in developing advanced aerospace systems at the lowest possible cost. If a design bureau fails to win its share of competitions, it is dissolved and the best engineers from this bureau are reassigned to the more productive bureaus; the nonproductive personnel from the dissolved bureau are assigned to menial engineering or teaching positions of little significance. Soviet emphasis on increasing the experience level of design teams and maintaining a low turnover rate in design personnel has attracted the attention of some individuals in the United States, and this is discussed in later sections. The Communist Party's stranglehold on all aspects of the Soviet academic-military-industrial complex, however, strongly resembles George Orwell's *1984*, where the state is the master of the individual from cradle to grave.

Almost all Russian engineers I have talked with appeared outwardly content with their role in life, perhaps because the Communist Party has decreed that science and technology must progress to its most advanced state, and educated persons are held in high esteem in the Soviet society. However, I could also detect a yearning in some individuals to be able to pursue their careers without Big Brother (i.e., the KGB) watching over their shoulders, telling them when they could travel outside the Soviet Union, who they can talk with, and what time they had to return to their hotels.

Another significant difference between Soviet and American design philosophy is that Soviet design bureaus frequently develop aerospace systems solely to advance the state-of-the-art, and the systems developed are not necessarily mass produced; this has saved the Russians billions of dollars in production costs. (I.e., because a weapon system is developed it does not mean that the weapon should be automatically mass produced; the need and cost-effectiveness of the system relative to other priorities must be taken into account.) Thus, while the United States waited almost two decades to develop another air superiority fighter aircraft, the F-15, the Soviets spent this period continually developing and flight-testing fighter aircraft to improve their technological capabilities in aircraft and engine design, but not all of the tested aircraft were mass produced. As is shown in later sections, this design philosophy has permitted the Soviets to build up a three to five year lead in fighter aircraft development and deployment over the United States.

In summary, Soviet experimental design bureaus are responsible for the design and development of ICBMs, space launch vehicles, spacecraft, aircraft, rocket and jet engines, and other significant systems for their space and defense establishment. These design bureaus (1) are staffed by experienced engineers who spend almost all of their productive lives working with each other, (2) perform very specialized work dictated by the Communist Party, and (3) are judged to be extremely competent by American engineers and intelligence analysts who have had the opportunity to examine Soviet hardware on a firsthand basis. It is Soviet policy to maintain the state-of-the-art by frequently developing prototype aerospace systems that are not necessarily mass produced. The successful Soviet design bureau system has been in existence for decades and will continue to play important roles in the future, especially with the advent of advanced aerospace systems that deploy both jet and rocket engines.

SAS-14

Whenever the Soviets embark on the research, design, and development of an advanced fighter aircraft or a bomber weapon system, the institute that commands the attention of the U.S. intelligence community is the Central Aerohydrodynamics Institute, which the Soviets abbreviate as TsAGI. The Central Aerohydrodynamics Institute, commonly pronounced "TSAH-gee" by intelligence specialists, has two facilities: its offices are located in Moscow, but its aerodynamic test facilities and wind tunnels are located outside of Moscow, near Zhukovskiy. TsAGI is one of a number of institutes and design bureaus that report to the Ministry of Aircraft Production, headed by P. V. Dimentiev, mentioned earlier. A U.S. intelligence analyst once described TsAGI as a combination of the Massachusetts Institute of Technology, Air Force Arnold Engineering Development Center in Tennessee, and the NASA Lewis Research Center in Ohio; this was another way of saying that TsAGI is not an institute in the generic sense, because it (1) is in fact extensively involved in the research, development, and testing of advanced fighter and bomber aircraft systems, (2) participates in the management of critical Soviet aerospace programs, and (3) is an adviser to the Ministry of Defense, Ministry of Aircraft Production, and the design bureaus affiliated with these ministries. Some departments at TsAGI are examining preliminary designs of advanced aircraft which can travel better than seven times the speed of sound, or roughly twice as fast as current operational interceptor aircraft in the U.S. inventory. While this effort is intended to support the development of advanced bomber weapons systems of the type that would revolutionize airpower, TsAGI was directed by Grechko's Ministry of Defense and Kirillin's State Committee for Science and Technology to coordinate its research work with Keldysh's Academy of Sciences in conjunction with the academy's space program, which is discussed in later sections. I had the opportunity to meet Vladimir Sychev, one of the deputy directors of TsAGI, in Argentina and a year later in Germany. Sychev, an enthusiast of Western movies (though he did not care for the film version of Dr. Zhivago), is also TsAGI's managerial and technical liaison man with the Academy of Sciences. As is shown later, the academy is involved in the development of a reusable spaceship that can operate like an airplane and routinely shuttle people, materials, and satellites into earth orbit and back again; TsAGI is responsible for developing the lower stage of this reusable space shuttle.

Another significant facility in the Soviet Union is the Moscow Aviation Institute (MAI), which prepares promising engineers and scientists for careers in the Soviet aerospace complex, and is very actively involved in the development of both rocket and jet engines, aircraft, and space systems. The Moscow Aviation Institute has very close ties with V. P. Glushko's rocket engine design bureau and is very much involved in the development of the upper stage of the Soviet space shuttle, which is discussed in later sections. Its importance in the Soviet aerospace establishment is exemplified by the caliber of personnel who work within its confines, and this includes Sergei P. Korolev, the mastermind of the Soviet space program until his death in 1966; Valentin P. Glushko, chief designer of rocket engines, U.S.S.R.; and Georgiy Petrov, director of the Institute of Space Research. Two prominent engineers I have met from the Moscow Aviation Institute are Vladimir Sosounov, who is also the head of a department within the Ministry of Aircraft Production (Sosounov demonstrated tremendous knowledge of U.S. aerospace programs, such as the Air Force's F-15 air superiority fighter), and Gennadi Dimentiev, a high-level Communist Party official and aerospace manager. As a rule, the Soviets go to great lengths to protect the identity and importance of its top space scientists and engineers, and while the reason they give is to protect these men from possible assassination from the West, the KGB likes to minimize the amount of information released to the West about classified programs and the personnel involved; Gennadi Dimentiev falls into this category and is mentioned in later sections.

A prestigious institute—sometimes called the MIT of the Soviet Union—is the Moscow Physical Technical Institute (MPTI), directed by red-haired and debonair Academician Oleg Belotserkovskiy. Top scientists from the Academy of Sciences teach at this institute on a part-time basis, and promising MPTI students spend some of their time at various Moscow-based design bureaus and the Central Aerohydro-dynamics Institute to receive on-the-job training. In what has been termed as "The Gogish Incident" in U.S. intelligence circles, I became involved in a heated and emotional exchange with Belotserkovskiy on the mezzanine floor of the Provincial Hotel in Mar del Plata, Argentina, when the academician's outburst in the presence of KGB watchdogs all but confirmed the existence of a viable space shuttle development program in the Soviet Union; scientists from MPTI are very much involved in the effort.

Academician Georgiy Petrov's Institute of Space Research—the NASA of the Soviet Union—is by far the most important space-related institute in Russia, and as is the case with TsAGI, this again demonstrates that Soviet institutes conduct significant research and manage technical programs that are of vital concern to the Communist Party. By comparison, most U.S. institutes are involved in regional, independent, noncoordinated projects, and they do not participate in the management of important national defense programs; the few institutes that were engaged in classified research for the Department of Defense, such as the MIT Instrumentation Laboratory, curtailed this activity after the violent 1971 anti-war campus demonstrations spearheaded by left-wing groups.

The Institute of Mechanical Problems of the U.S.S.R. Academy of Sciences is highly regarded by most experts in the mechanical engineering field. One significant program at the institute concerns the effects of laser beams on various target materials. This work has been related to the potential use of laser weapons in orbital space, which is discussed in later sections. Part-time KGB specialist Yuri Riazantsev is affiliated with this institute.

The Institute for Machine Studies has the notorious reputation of graduating engineering candidates for the KGB. Igor Prissevok, the junior-level KGB officer who accompanies Soviet cosmonauts abroad and was responsible for preventing the defection of Soviet scientists at the October 1972 International Astronautical Congress in Vienna, Austria, graduated from this institute.

The Institute of Automatics and Telemechanics of the U.S.S.R. Academy of Sciences, formerly headed by Vadim Trapeznikov, who is now a member of the Central Committee of the Communist Party and assistant to both Keldysh and Kirillin of the academy and State Committee for Science and Technology, is the center for Soviet computer and automatic control theory and applications. This institute, currently headed by Academician Boris V. Petrov (note: not to be confused with Academician Georgiy Petrov of the Institute of Space Research), has contributed significantly to the design and management of (1) automatic control and guidance systems of Soviet space probes to the moon, Mars, and Venus, (2) manned and unmanned orbital spacecraft, and (3) advanced manned aircraft and spacecraft systems. Boris Petrov, who had a heart attack in August 1971, is usually the official Soviet spokesman quoted by the news media following a major Soviet space launch. He was also the head of the Soviet delegation during the recent U.S.-U.S.S.R. discussions on the joint space mission planned for 1975. Petrov's deputy, Georgiy M. Ulanov, a large fatherly scientist, represented the Russians during the preliminary negotiations with NASA space officials on the joint mission in October 1970. I found Boris Petrov and Georgiy Ulanov to be congenial and approachable individuals, though both were exceptionally careful during technical discussions and were thoroughly briefed on the risks of making an inadvertent disclosure that might compromise the security of the classified programs they worked on.

The Institute of Mechanics of Moscow State University has a reputation for its rocket engine and research studies that are applicable to advanced bomber aircraft and the Soviet space shuttle. Its director is corresponding (non-voting) member of the U.S.S.R. Academy of Sciences Gorimir G. Chernyi, an aggressive friendly researcher who commands the respect of his subordinates and is known in private Soviet circles as the gypsy, not because he is one of the few Soviets who is permitted to travel extensively in the West, but because he is in fact a gypsy. In my last exchange with Chernyi in 1970, the scientist exhibited an insatiable appetite for any information about U.S. rocket engine developments.

The aforementioned institutes are only a sampling of the hundreds of institutes in the Soviet Union that contribute to their space and defense effort, but they are among the most important. Plate 12 shows the significant administrative organs and scientific and technical establishments in the Soviet Union discussed in chapters one and three. It should be reiterated that this complex is (a) literally controlled by a handful of men in the Politburo of the Central Committee of the Communist Party and (b) can be credited with catapulting the Soviet Union from a Stone Age third-rate country to a leading world and space power.

With this background, it is possible to discuss in meaningful terms the published and unpublished aspects of the Soviet space program, their military strategy, weapons systems, and future plans.

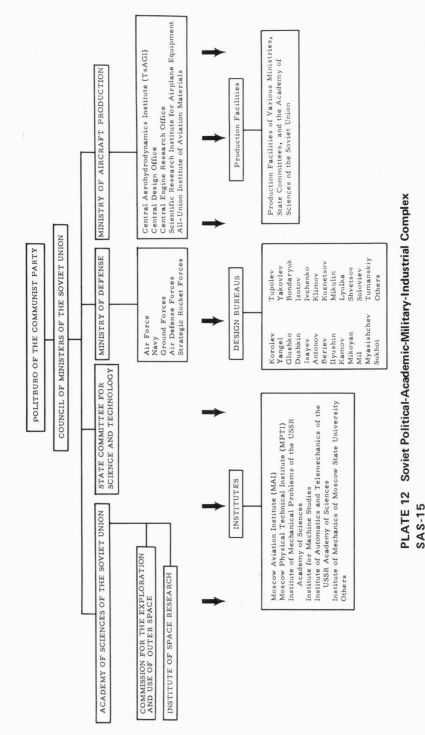

PLATE 12 Soviet Political-Academic-Military-Industrial Complex

SAS-15

Chapter 5

Soviet
Rocket Engines

The analysis of the Soviet space effort by pseudo experts has created a great gulf between what has been reported and fact. This has had a detrimental effect on the U.S. space program, public opinion (as demonstrated by the low-keyed coverage of the Apollo 17 mission after the midnight launch of the Saturn V moon rocket), and intelligence analysis techniques. The discussion below begins with the Soviet rocket engine development programs, and comparisons are made with the U.S. effort to place the analysis in proper prospective.

During the 1967 and 1969 Paris Air Shows and other space exhibits throughout the world, the Soviets displayed for the first time four of its liquid propellant rocket engines. Prior to these public displays the majority of the intelligence analysis performed on Soviet rocket systems was based on raw intelligence reports; CIA analysts will be among the first to concede that there is a world of difference between an intelligence assessment based on fragmentary word-of-mouth information about a rocket engine, and that based on visual examination of the rocket engine itself. The rocket engines displayed by the Russians were designated as follows (see Plate 13).

(1) RD-107 The rocket engine used in the booster stages of the Soviet first generation ICBM and standard launch vehicle used to inject Soviet cosmonauts into orbit.

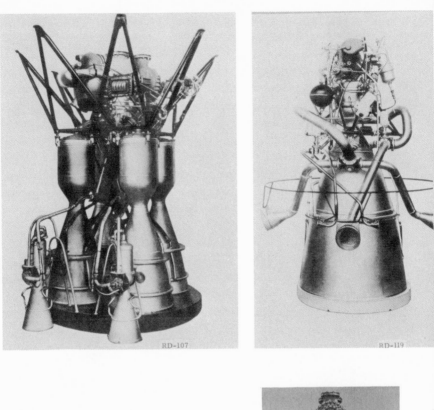

PLATE 13   Exhibited Russian Rocket Engines

(2) RD-119 The rocket engine used in the upper stage of the Cosmos launch vehicle used to inject small payloads into earth orbit.

(3) Molniya: The rocket engine used in the Soviet orbiting communications satellite.

(4) RD-214 The rocket engine used in the lower stage of the Cosmos launch vehicle used to inject small payloads into earth orbit.

Whereas some analysts who should have known better adamantly took the position that the Russians were using exotic propellants and advanced rockets to support their space effort, they found instead that the Russians achieved the greatest coup of the century, the launching of the world's first satellite using rocket engines that in the words of one analyst were: " ... crude, simple, and looked as if they were designed by a blacksmith." An evaluation of the aforementioned rocket engines revealed that Russian designers placed a great amount of emphasis on the simplicity of engine designs, with the overall objective being to increase reliability and reduce costs; they sacrificed rocket engine performance and efficiency by not designing sophisticated, complicated, advanced, and costly engines. This was a significant revelation because after the launch of Sputnik, the United States took the opposite approach: stringent and costly measures were undertaken to produce rocket engines that delivered the highest possible performance; U.S. government specifications required contractors to design and develop rocket engines that were of minimum weight and volume because numerous studies showed that these were desirable features in rocket systems. To meet the required design specifications, U.S. contractors developed new materials and tested rocket engine concepts that were sometimes exotic modifications of existing designs. In addition to being cheaper than American engines and giving the outward appearance that they were designed in an amateur rocket club's basement, the Russian rocket engines had distinct features that were generally different from U.S. designs: The RD-107 rocket engine, deployed in the first stage of the rocket that was used to launch Sputnik in 1957 and Cosmonaut Yuri Gagarin in 1961, consists of four combustion chambers (cylindrical cavities where propellants are mixed and burned) and four exhaust nozzles (conical sections connected to the combustion chambers so that the combustion exhaust products can be ejected into the atmosphere to produce thrust), while most U.S. designs employ one combustion chamber and one connecting exhaust nozzle for each engine. This difference becomes noticeable to the casual observer by comparing launch vehicles: The Saturn V Apollo moon rocket used five main rocket engines in the first stage, and these were discernible by the five protruding exhaust nozzles at the bottom of the stage; the Yuri Gagarin launch vehicle used five RD-107 type rocket engines, but to those who first saw it at the 1967 Paris Air Show, it was a ghastly sight—twenty main exhaust nozzles were exposed. This launch vehicle is discussed in later paragraphs, but the conclusion that has been corroborated by reliable sources is that in the early 1950s, the Russians lacked the required test, manufacturing, and production facilities to develop large rocket components, so they did the next best thing: instead of designing a single unit that could deliver about 225,000 pounds thrust, the Russians built four smaller units that could deliver one-fourth the required thrust each, and they developed the hardware to operate the four units as a single engine—the RD-107.

In another basic design difference, U.S. designers built their rocket engines with a gimbal capability, which meant that the engines could be pivoted (gimballed) during flight from their normal thrust direction to permit the on-board computers to steer the vehicle along a predetermined or acceptable trajectory. However, the Russian space displays showed that their rocket engines are rigidly fastened to the launch vehicle and gimballing is achieved by using additional smaller engines called vernier rocket engines, which are used exclusively to steer the rocket. While American engineers have often remarked that the Russian solutions to large thrust rocket engines and gimballing lacked originality, were crude, and added unnecessary weight to the launch system, the Russian design engineers that I talked with maintained that the systems were simple, reliable, durable, and got the job done.

To place the aforementioned comments in their proper prospective, it should be noted that the RD-107 and RD-214 rocket engines were designed during the middle 1950s and were based on the technology that the captured German scientists brought to Russia after World War II and the expertise of Russia's own scientists, whose serious rocket efforts can be traced back to the early 1900s. The interesting point is that these rocket engines, developed almost a quarter of a century ago, are still being used by the Soviets to support both their manned and unmanned space programs. Though they have been described as inefficient by Western space experts, they have greatly contributed to the Russian space effort and they have served their purpose many times over.

The questions that experts have recently asked are: if the Russians have had success with their space program using crude space systems developed during the middle 1950s, what could they have done with technologically advanced aerospace systems? If it has been almost twenty years since the Russians developed the RD-107 and RD-214 rocket engines, what have they been doing since then? How does their work compare to the U.S. effort? To answer these questions, one must have a basic understanding of advanced rocket work in the United States.

The most advanced liquid propellant rocket engine currently being developed in the West is called the Space Shuttle Main Engine and this work is being done by Rocketdyne. When development work on the engine is completed, it will be used in the upper stage of the United States' space shuttle launch vehicle. The U.S. space shuttle, currently being developed by North American Rockwell for the National Aeronautics and Space Administration (NASA), is a reusable two-stage launch vehicle that should become operational by 1980. It will be capable of delivering men, equipment, satellites, and other space payloads into earth orbit. The shuttle differs from other space launch vehicles developed by the United States because it can be returned to earth and used over again, perhaps as many as 500 times. It will be cheaper to use the space shuttle because it becomes an expensive proposition to keep throwing away launch vehicles after each use, just as it would be expensive to discard a jetliner after each transcontinental flight. The first stage of the U.S. space shuttle is unmanned and consists of two giant solid rocket motors (i.e., the propellant components are in a solid state instead of a liquid state, like that of the Saturn V moon rocket). After launch from Cape Kennedy, these motors will land in the Atlantic Ocean where they can be recovered and refurbished for use over again. The manned upper stage of the space shuttle is both an airplane and an advanced rocket powered spacecraft, and is about the size of a medium-range jetliner such as the Boeing 727. Three Rocketdyne rocket engines (i.e., the space shuttle main engines), designed to deliver almost half a million pounds thrust each, will power the second stage of the shuttle into orbit, and for that reason the second stage is also called the orbiter stage. After completing its mission, perhaps to deliver or retrieve a satellite, the orbiter stage will reenter the atmosphere and land at an airfield like an airplane. One of the keys to the success of the U.S. space shuttle effort is the development of rocket engines that can be used over again and deliver higher performance than rocket engines developed in the past; a nation that embarks on a space shuttle program must have the capability to develop these advanced rocket engines. In the United States, through the foresight of the Air Force and NASA, this technology was developed during the middle and late 1960s, and Rocketdyne, the builder of the moon rocket engines, was selected by NASA in 1971 to build the shuttle's rocket engines.

A thorough evaluation of the space shuttle rocket engine requirements by U.S. experts indicated that this engine, supposedly the most advanced engine in the world, would have to (1) provide about 6 percent higher performance than the upper stage engine of the Saturn V moon rocket and (2) be used up to 100 times before being discarded. (Note: due to severe thermal requirements, the reusability of rocket engines, i.e., the propulsion system, is less than that for the airframe of the space shuttle.) To achieve these design objectives, U.S. engineers selected liquid oxygen and hydrogen propellants. Liquid oxygen is called the oxidizer and liquid hydrogen is called the fuel; both must be mixed and ignited in the combustion chamber and exhausted through a nozzle to produce thrust. To be used in a rocket engine, oxygen and hydrogen must be stored in propellant tanks in a liquid state at hundreds of degrees below zero. If adequate storage conditions are not provided for, these propellant components, which are sometimes called cryogenics because of their low temperature characteristics, can boil off like water in a kettle. Cryogenic propellants are desirable to use in advanced rocket engines because they offer higher performance than most other propellant combinations. Detailed U.S. studies showed that the required space shuttle mission performance levels could not be attained unless oxygen-hydrogen is combusted in rocket combustion chambers under extreme pressures of about 200 times greater than atmospheric pressure (i.e., about 3,000 pounds per square inch). Relative to past engine designs, these are ambitious design requirements; the rocket engines in the upper stages of the Saturn V moon rocket, for example, combust their propellants at pressures of about 53 times greater than atmospheric pressure. Because of the high combustion pressures required for the space shuttle main engine, U.S. designers began calling this type of engine the high chamber pressure rocket engine.

During the middle and late 1960s, I had the opportunity to conduct numerous mission studies on advanced space shuttle-type rocket systems using high chamber pressure rocket engines, and I became familiar with the problems involved in integrating this type of propulsion system with advanced reusable space vehicles. Because I was also program manager of foreign technology at Pratt & Whitney's Florida Research and Development Center, I was in a position to look for developments in Soviet rocketry that would indicate that they too were developing a high chamber pressure rocket engine. The first sign came through normal analysis of the engine chamber pressure values of exhibited rocket engines from both countries. The supposedly obsolete Russian RD-107 and RD-119 rocket engines combusted their propellants at chamber pressure values that were 60 and 80 times greater than atmospheric pressure—much higher than American rocket engines built during the same time period. An examination of textbooks approved by the KGB showed that Soviet rocket engine designers advocated the development of high chamber pressure rocket engines, but for security reasons they were careful not to discuss Russian engines developed during the middle-late 1960s.

During the 1969 Paris Air Show a Soviet design engineer told me that the Soviets were conducting detailed analytical studies (i.e., paper studies) on oxygen-hydrogen high chamber pressure rocket engines, and the intended application of the engine was an aerospace system that the Russians called the Rocketoplan—the orbiter stage of the Russian space shuttle, which looks like an airplane and is powered by rocket engines. I reported this and related information to my CIA contact when I returned to the States. Civilian intelligence analysts with CIA connections also heard about the Russian disclosure in Paris, and they expressed hope that I could follow through and learn more about Soviet rocket engine and space shuttle plans at a space conference that was scheduled for October 1969 in Argentina.

At the Argentine meeting I learned that the Russian high chamber pressure rocket engine was being developed by V. P. Glushko's design bureau. The Russian engine would operate at chamber pressure values that would be about 15 percent higher than the U.S. space shuttle main rocket engine.

With the Russians developing a space shuttle-type rocket engine and the U.S. similarly expending millions of dollars in the development of such an engine, a unique opportunity presented itself in Konstanz, Germany, in October 1970 when I met Professor Gennadi Dimentiev of the Moscow Aviation Institute. Though Dimentiev was the mystery member of the Soviet delegation to the Mar del Plata and Konstanz international scientific meetings, I was informed by a reliable source that Dimentiev was using a pseudonym to conceal the fact that he is a high-level member of the Soviet political establishment and one of the heads of the Soviet space program. In a very private meeting Dimentiev and I sat at a secluded table overlooking the Lake Bodensee, and with KGB scientist Yuri Riazantsev translating, I proposed the exchange of aerospace hardware between the United States and the Soviet Union with the overall objective being to reduce aerospace expenditures between both countries. In this discussion, Dimentiev acknowledged that he knew V. P. Glushko personally and thought that the proposal had merit because the Russians also found that the development cost of their high chamber pressure rocket engine was enormous. He suggested that I contact the U.S. State Department and make the proposal with official U.S. government backing. He said that he would talk with Glushko and assured me that the Soviet scientific community would support the proposal before their government. Additionally, he stated that the Russians would be willing to declassify their high chamber pressure rocket engine program, as was done by the United States only months earlier in 1970, to participate in a preliminary exchange of engine hardware with the United States.

When I returned to the United States, I summarized these discussions in another report for the CIA. After one month I was contacted by the agency and given the name of a U.S. government official who could pursue the proposal through overt government channels. After three months the proposal was rejected by the U.S. State Department's Office of Munitions Control and the matter was closed. Though there were pros and cons to the technical exchange, this futile effort did establish that the Soviets were aggressively pursuing the development of a high chamber pressure rocket engine for use in their space shuttle. For security and other reasons the details of the exchange and the existence of the Russian effort were never made public.

It can be stated that the Soviet liquid propellant rocket engines developed during the 1950s were simple, rugged, and reliable, and are currently being used in support of their manned and unmanned space program. During the past fifteen years the Soviets have significantly improved the state-of-the-art of their rocket technology and are currently developing an oxygen-hydrogen high chamber pressure rocket engine that will be used in the orbiter stage of their space shuttle. Analysis and reliable information indicates that (1) the Soviet space shuttle engine will deliver slightly higher performance than the shuttle engine being developed by Rocketdyne and (2) V. P. Glushko's rocket engine design bureau has been assured continuous support by the Central Committee of the Communist Party and U.S.S.R. Academy of Sciences to develop this and other advanced engines. In general, the Soviet liquid propellant rocket engine effort is first-rate and can support an aggressive space program for the 1970s and 1980s.

SAS-16

Chapter 6

Soviet Propellants

Analysis of rocket propellants used by the Soviets during the 1950s and 1960s surprised many U.S. space engineers. In the United States the emphasis has always been on using propellants that can deliver the highest possible performance, and this meant the use of cryogenic rocket propellants and the research and testing of exotic propellant combinations for future space systems. As mentioned in the previous chapter, the most advanced operational rocket engines in the U.S. inventory use oxygen-hydrogen propellants; this includes the upper stages of the Saturn V moon rocket and the Centaur stage of the Atlas/Centaur launch vehicle, which launched the unmanned Surveyor moon landing craft and the Mariner Mars orbiter. A propellant combination that has been frequently tested in the U.S. is fluorine-hydrogen. Liquid fluorine is another supercold cryogenic propellant component, but it is very toxic and when contaminated with foreign matter it frequently burns out of control; test engineers will be the first to testify that fluorine explosions are among the worst. U.S. engineers have been willing to risk testing fluorine-hydrogen rocket engines because these types of engines can deliver about 5 percent higher performance than existing operational oxygen-hydrogen rocket engines. The Russian propellant selection philosophy, however, has been very conservative. For example, no known Russian oxygen-hydrogen rocket engine has ever flown in a vehicle launched during the 1950s and 1960s. Instead, the Russians have relied on simpler and lower-performing propellant combinations such as oxygen-dimethyl-hydrazine, oxygen-kerosene, and nitric acid-kerosene, whose operating characteristics have been understood for decades, though to the layman the names of propellant components are sometimes tongue twisters. Soviet research on fluorine-type propulsion systems has been limited due to explosions in their test laboratories and their belief that fluorine should never be used in launch vehicles because it could dangerously contaminate the atmosphere, especially during abort situations when propellant must be dumped overboard. For manned spaceflight missions, Soviet designers believe that fluorine would pose an unnecessary hazard to their cosmonauts because of its toxic, corrosive, and explosive characteristics.

Whereas numerous classified U.S. studies have been conducted on fluorine propellant systems for use exclusively in orbital space, again because of our emphasis on maximum performance systems, one of the chiefs of the Soviet space program told me that the Soviets consider reliability and the storability of propellants in space infinitely more important than performance. His logic was that space systems should be built for long term use, and this is possible only if the propulsion systems used are reliable and can operate in space for months without being serviced. According to a prominent spacecraft designer from the Moscow Aviation Institute, the United States has not studied the problems of orbital spacecraft systems thoroughly. The Russian correctly noted that the United States places too much emphasis on engine performance rather than system performance (i.e., the system includes rocket engines, propellant tanks, body structure, command module, payload, etc.); it is possible to design non-cryogenic spacecraft systems that can perform better than cryogenic spacecraft systems. The Russian was referring to orbital refueling operations and modular spacecraft construction, which are discussed in later sections, and are high on the Soviet priority list. One of his main objections to the use of cryogenics in orbital space is that for the programs the Soviets have planned, they would have to spend a tremendous amount of money on cryogenic space technology to cope with problems such as boil-off losses. (Cryogenic propellants can boil off like water in a kettle if adequate storage measures are not taken. In orbital space the problem becomes even more complex, especially for space missions that require extensive refueling operations and no servicing for months.)

The Soviet preference to use low-cost storable propellants such as nitric acid, nitrogen tetroxide, kerosene, hydrazine, and dimethyl-hydrazine—propellants that can be easily stored at room temperature without fear of boil-off problems—was underscored by a significant disclosure in Paris in 1969 by a Soviet space engineer. During a lengthy discussion on orbital spacecraft systems, the engineer revealed that the Soviets successfully simulated a space refueling operation by transferring nitric acid and kerosene propellants from one spacecraft to another during a Cosmos orbital space mission. This was an important disclosure because the Russian propellant transfer experiment was never publicized, yet refueling in space is a major Soviet objective. Detailed Russian mission studies have concluded that the capability to refuel spacecraft in orbit is more important than developing costly high performing orbital rocket propulsion systems. This is another way of saying that even though a Volkswagen can travel about ten miles more on a gallon of gasoline than a Chevrolet because the Volkswagen burns gasoline more efficiently, the flexibility and range of both cars could be greatly enhanced if they are permitted to refuel at a gasoline station. Just as gasoline stations are more important than the make of an automobile when it comes to driving across the United States, the Russians believe that refueling spacecraft in orbit is more important than the performance of the spacecraft. Because the Soviets are willing to settle for lower performing and easier-to-handle earth-storable propellants in orbital spacecraft, their decision also has military overtones: all current U.S. and U.S.S.R. ICBMs use earth-storable propellants because military weapons systems must be capable of being launched on a moment's notice, even though they are stored in their silos for months on end; supercold cryogenic propellants continuously boil off and must be replenished up to the time of launch, and this requires a large amount of manpower, support equipment, and time, which is not acceptable to the military. Unless certain costly technical problems are solved, cryogenic propellants are not conducive to long-duration military space missions that require a quick reaction time.

In summary, the Soviet space program has utilized simple earth-storable propellants such as nitric acid, nitrogen tetroxide, kerosene, and dimethylhydrazine, and on occasion, cryogenic oxygen. The technology and handling characteristics of these combinations have been understood by the Soviets for decades. In modern times, the Soviets continue to shy away from exotic propellant components such as fluorine, which is toxic and dangerous to work with. For future launch vehicles such as the upper stage of the Soviet space shuttle (the Rocketoplan, discussed in later sections), the Soviets will use oxygen-hydrogen (cryogenic) propellants. However, for systems that will operate entirely in space (The Rocketoplan is excluded from this category because it returns to earth for use again) the Soviets will develop reusable propulsion systems that burn earth-storable propellants; this decision is based on Soviet military considerations, emphasis on long-term propellant storability, ease of transferring propellants in space (relative to cryogenic propellants), and factors outlined in later sections. While American experts have recommended the use of cryogenic propellants for future orbital spacecraft, Soviet designers believe that earth-storable propellants are the key to developing a low-cost civilian and military orbital space capability. **SAS-17**

Chapter 7

Soviet Launch Vehicles

The backbone of a nation's space capability is its current and planned launch vehicle inventory. It took U.S. intelligence analysts almost a decade to conclude that the Soviets outplanned the United States in the development of space launch vehicles, and when the pieces to the puzzle were put together it was found that the Soviets came within an eyelash of returning lunar soil to the earth before the United States. For the most part the public and some government officials never knew how close the moon race really was. Furthermore, while the United States has been winding down its space program after the tremendous Apollo expeditions to the moon, it was determined that the Soviets have been secretly developing the potential to exploit the scientific and military uses of space through the utilization of their current launch vehicle inventory and the development of very advanced systems which have yet to be revealed. The paragraphs below summarize this effort.

The rocketship that the Soviets have used to launch their cosmonauts into orbit—sometimes referred to as the standard launch vehicle because of its application to numerous other space missions—was first displayed to the West during the 1967 Paris Air Show at Le Bourget. The standard launch vehicle is noticeably different from U.S. launch vehicles. The first stage, also called the booster, uses five main rocket engines (i.e., the RD-107 type engine discussed earlier), and since each engine consists of four exhaust nozzles, the aft end of the rocketship looks cluttered with nozzles. In addition to the twenty main rocket engine nozzles (i.e., five engines times four nozzles per engine) the booster stage has twelve smaller vernier rocket engines for steering the rocket. Therefore, as Plate 14 shows, there are thirty-two nozzles in the booster stage.

The other unusual feature of the booster is that it has a core stage surrounded by four strap-on modules that resemble elongated cones. (Also see the strap-ons in Plate 15, which shows the standard launch vehicle after lift-off.) When the propellant in the strap-ons is depleted the modules separate from the rocketship, much like a ladder swinging away from a building, permitting the core stage to continue to burn. In the early Sputnik launches the core stage was injected into orbit with the satellite and both circled the earth.

The standard launch vehicle should be considered to be one of the most outstanding planning achievements of the space age—a point that has not been made clear to the American public. The rocketship was the creation of the mastermind of the Soviet space program, Dr. Sergei P. Korolev, the brilliant scientist mentioned earlier who died unexpectedly in 1966 of stomach cancer. The booster stage was on the drawing boards during the early 1950s, it was developed in secret during the early-middle 1950s, it was deployed as a first generation ICBM immediately thereafter, and it was used on October 4, 1957 to launch sputnik. After the launch of the first sputniks, the booster was mated to various upper stages to launch both unmanned probes to the moon, Mars, and Venus, and manned spacecraft systems—Vostok, Voskhod, and Soyuz (discussed in later paragraphs). The magnitude of its versatility and usefulness can best be understood in terms of the U.S. launch vehicle inventory. In the 1950s, when the United States was developing launch vehicles that could deliver a few pounds of instruments into earth orbit, such as the three-pound Vanguard payload, Sergei Korolev's design teams were designing a standard launch vehicle that could deliver thousands of pounds of payload into orbit. Thus, when the Soviets launched three classes of manned spaceships in the 1960s (Vostok, Voskhod, and Soyuz), they could launch them all with the standard launch vehicle with slight modifications to its upper stage. On the other hand, during this period the United States built three manned spacecraft systems (Mercury, Gemini, and Apollo), but used the Redstone, Atlas, Titan-II, Saturn 1B, and Saturn V vehicles to launch them. Except for the Saturn V moon

PLATE 14 Russian Standard Launch Vehicle Being Transported to Launching Pad (Novosti Press Agency Photo)

PLATE 15 Russian Standard Launch Vehicle After Lift-Off (Novosti Press Agency Photo)

rocket, which was designed specifically to support the Apollo moon landing, the United States used four different classes of rocketships to launch its astronauts, compared to one for the Soviets. It was clear to many space experts that the Soviets planned better than the United States and developed a versatile launch vehicle that could accommodate numerous manned and unmanned space missions with the minimum use of new hardware. The significant point worth noting is that the standard launch vehicle, designed during the early 1950s and used in hundreds of space launches to date, will be used by the Soviets to support their space program in the 1970s. By comparison, most manned U.S. rocketships built have already flown into space, the production facilities for the Saturn series of launch vehicles have been closed, the last Apollo moon mission was flown in December 1972, and except for the Skylab space station missions in 1973 and the joint U.S.-U.S.S.R. space mission planned for 1975 the United States' manned spaceflight program will be dormant until about 1980, when the space shuttle is supposed to become fully operational.

A final comment is in order about the "old" standard launch vehicle. It was once described by an intelligence analyst as "a workhorse booster of battleship construction that is lousy by American standards, but it works." This was another way of saying that Soviet designers stress rocket durability and reliability, instead of minimum weight and maximum performance. The experience of the past twenty years has shown that the Russians emphasize planning first and technology afterwards; in the United States, because of our affluence, tolerance to do things over again, and the turnover of space managers, we tend to emphasize technology first and planning afterwards; we have thousands of engineers who enjoy challenges, particularly developing the most sophisticated, lightweight, and minimum volume hardware that can be tested over and over again in the most elaborate facilities, because some space managers and planners who should know better want it done this way. It explains why a beer bottle thrown against an unpressurized Centaur stage (i.e., the upper stage of the launch vehicle used for the U.S. Surveyor and Pioneer space missions) might do irreparable damage, while the effects of a cannonball thrown against the Russian standard launch vehicle would hardly be noticeable. Again, it explains why the NASA tour guide at the Cape Kennedy launch complex shows the "progress" of the U.S. space program in the form of obsolete launch pads that resemble the oil derricks of a dilapidated ghost town, while the launch complex for the standard launch vehicle at Tyuratam—the Cape Kennedy of the U.S.S.R.—is still operational.

> In summary, the Russians built a durable, low-cost workhorse standard launch vehicle in the 1950s that can deliver up to 16,500 pounds of useful payload into orbit, and it will continue to be used to support their manned and unmanned space program in the 1970s. This vehicle cannot be considered an advanced technological achievement by American standards, but it must be regarded as an outstanding planning achievement and an asset to the Soviet space program.
>
> **SAS-18**

The next largest size space launcher that the Soviets developed was the Proton launch vehicle. It was first used in 1965 to inject the 27,000 pound Proton 1 scientific research satellite into earth orbit. Just as Sputnik was launched with the booster stage of the standard launch vehicle, the Proton satellite was launched using only the booster stage of the Proton launch vehicle. With the subsequent addition of upper stages, the Soviets increased the lift capability of the vehicle; the 37,000 pound Proton 4 satellite was launched in 1968, and the 40,000 pound Salyut space station was launched in 1971. This launch vehicle has also been used to support (1) the unmanned circumlunar Zond missions that photographed the lunar surface during a flyby and returned to earth, (2) the unmanned Luna missions that explored the lunar surface by automatic means with a lunar rover vehicle (Lunokhod), (3) the unmanned Luna missions that returned several ounces of lunar soil to earth by automatic means, and (4) the unmanned planetary probes to Mars and Venus. With further improvements of its upper stages the Proton booster will be capable of delivering about 60,000 pounds of useful payload into an earth orbit. In other words, the Soviets again developed a launch vehicle that could accommodate a large variety of space missions; with a payload capacity of 27,000 to about 60,000 pounds, depending on the arrangement of the upper stages, the Soviets achieved with the Proton launch vehicle something that the United States could not achieve with three launch vehicles: the United States' Saturn 1 and Saturn 1B (NASA launch vehicles that preceded the Saturn V moon rocket) and Titan III-C (Air Force launch vehicle) were designed to deliver 22,000, 37,000, and 25,000 pounds payload into earth orbit respectively. The payload capacities of these launch systems were not only similar, but each vehicle was designed with a very narrow payload range; they were not versatile for use in widely different missions (i.e., earth orbit, lunar, and planetary missions). These design limitations also created a tremendous payload gap in the U.S. launch vehicle inventory; the lift capacity of the three launch vehicles was in the 22,000 to 37,000 pound range, while that of the Saturn V moon rocket was 280,000 pounds payload to earth orbit. Therefore, all U.S. payload systems had to weigh less than 37,000 pounds because no organization could justify launching a satellite with the $185 million Saturn V launch system.

The Soviets have never displayed the Proton launch vehicle in the West or publicly in the Soviet Union. Therefore, unclassified Western data published on this system are at best based on speculation or the analysis of Soviet (KGB-approved) press releases and existing factual data about the Soviet space program. But it is known that the Proton launch vehicle is about half as large as the Saturn V moon rocket. It was once described by a Soviet design engineer, who had too much to drink, as follows: "It has a super high thrust center (core) stage and is surrounded by many, many lower thrust strap-on rockets that burn out before the center stage." This was his way of saying that the vehicle uses at least four strap-on modules and has a core stage that is powered by new higher thrust rocket engines. In a discussion with another Soviet designer, I learned that these rocket engines operate at chamber pressure values that are 100 times greater than atmospheric pressure, or roughly twice as high as the rocket engines of NASA's Saturn 1B, which were built about the same time period. In other words, the Russians were still designing higher chamber pressure rocket engines than the United States.

In contrast to the designs of the upper stages of U.S. launch configurations (i.e., Atlas/Centaur, Saturn 1, Saturn 1B, and Saturn V moon rockets) the upper stages of the Proton launch vehicle do not use hydrogen (cryogenic) fuel.

In summary, the Soviets built another durable and versatile launch vehicle capable of delivering between 27,000 and ultimately 60,000 pounds of payload into earth orbit. The Proton launch vehicle has been used for near-earth, lunar, and planetary space missions, and it will continue to be used in the 1970s with improved

With the Proton launch vehicle operational in 1965, it seemed logical to most Sovietologists that the Soviets would be working on a larger system. In a search that spanned three continents, and through discussions with Soviet designers and the heads of their space effort, I acquired bits and pieces of information on this intriguing, yet elusive, advanced launch vehicle. The story began in 1965 in Athens, Greece, when Academician G. L. Grodzovskiy photographed the design characteristics of the Saturn V and its test facilities from slides shown by Dr. Wernher von Braun during his presentation on the status of the Apollo manned spaceflight program at the XVI International Astronautical Congress. (Academician Grodzovskiy is one of the secret heads of the Soviet space program.) In Athens, photographer Grodzovskiy was not only fulfilling an overt KGB intelligence requirement, but he wanted to personally present the Saturn V data to Academician V. P. Glushko and other members of the Soviet space team, who were finalizing the design of their own moon rocket. Two years later at another international meeting in Belgrade, Yugoslavia, a reliable Eastern bloc source told me about significant Soviet work on advanced boosters. Among other things, he said that the Soviets were developing a launch vehicle that was larger than the Saturn V moon rocket, and a full-scale prototype version of the launch system was being prepared for testing; this launch vehicle, subsequently called the Super Booster by U.S. space experts, would meet the payload requirements of the Soviet space program for the next ten years. At this meeting Academicians G. L. Grodzovskiy and O. M. Belotserkovskiy were asked about the Soviet manned lunar landing timetable. In a stuffy room of the Youth House on Makedonska Street in Belgrade, the academicians jointly stated that the Soviets would land men on the moon about the same time as Apollo. Their statements were significant because they not only confirmed that the Soviets did not withdraw from the moon race as some Western observers believed, but they indicated that their moon rocket—the Super Booster—was indeed ready for its first test flight.

In early 1968, however, the Super Booster's test flight was cancelled when the Soviets encountered unexpected problems with the first stage's rocket engines. During this period, intensive work was also being conducted on a system that could be launched by the Proton launch vehicle and would (1) land an instrumented package on the lunar surface, (2) retrieve several ounces of lunar soil, and (3) return to earth automatically. Because of the design problems with the Super Booster, the unmanned lunar soil retrieval effort was given A-1 priority.

In May 1969, two months before the Apollo 11 moon launch, I participated in a reception attended by Russian scientists at the Europa Hotel in Venice and learned that the design problem with the Super Booster had been resolved and the vehicle was resting on the launch pad. In an effort to determine whether the Soviets had anything planned before or during the Apollo 11 mission, I confronted KGB Colonel Nikolai Beloussov—the mastermind of the Christine Keeler-John Profumo sex scandal and high-level international negotiator for the Soviets—and commented that it appeared that the United States would beat the Soviets to the moon. I was looking for his reaction more than anything else. Beloussov gazed at the ceiling with a drink in his hand and replied, "You may be surprised!" I reported this and other information to the CIA, and noted that the Russians had plans for some type of moon launch before Apollo 11. Other reports came in as well, and the U.S. intelligence community was prepared for the Russian launch. Numerous intelligence analysts asked: How can the Russians beat us to the moon if they have not flown their moon rocket or conducted orbital maneuvers with new hardware (as was done with the Apollo 7 through 10 missions)? One theory offered, and with each passing day it became obvious that this was not just a theory, was that the Soviet moon rocket was larger and more advanced than the Saturn V, but was not ready because the Russians were still trying to get all the bugs out of it. However, their automated Luna spacecraft, which was designed to retrieve lunar soil and automatically blast off from the moon for an earth recovery, could still be launched before Apollo 11, though it had not been thoroughly tested. Because the reliability of the Proton launch vehicle was low compared to the standard launch vehicle, and the Super Booster had yet to fly, the Soviet manned lunar mission plan in 1969 included the launch of cosmonauts in the reliable standard launch vehicle and the transfer of these cosmonauts to their moon spaceship, already in earth orbit. In other words, the Soviets did not want to risk launching their moon rocket with men. Since the Soviets had already successfully sent unmanned Zond spaceships around the moon and back to earth in late 1968, it was accepted by almost all experts that the Soviets could send men around the moon and back if they chose to. The Soviet secret plan was to use the Super Booster to launch an unmanned Zond-type spacecraft and its accompanying circumlunar propulsion system, complete with prototype hardware and mock-up lunar stages, into earth orbit. A few hours later, Soviet cosmonauts would be launched with the standard launch vehicle. The cosmonauts would then transfer to the Zond-type spacecraft with its connecting circumlunar propulsion system in earth orbit using the crew transfer techniques developed only months earlier. In January 1969, the Soviets transferred cosmonauts from Soyuz 5 to Soyuz 4 after these spaceships rendezvoused and docked. The Soviets publicized the Soyuz 4 and Soyuz 5 crew transfer experiment as relating to the establishment of orbiting space stations and cosmonaut space rescue operations, but they made no mention that the mission was also the prelude to the transfer of cosmonauts from a Soyuz spaceship to a moonship. Thus, the Soviet plan was to send cosmonauts on a circumlunar trajectory before Apollo 11, so they could test their lunar hardware and simulate a future manned lunar landing from the safety of a manned spaceship that would loop around the moon and return to earth. With the cosmonauts in the vicinity of the moon conducting tests, the unmanned Luna spacecraft launched by the Proton vehicle a few days earlier would attempt to soft-land, pick up several ounces of lunar soil, and return to earth. If all aspects of both missions were successful, the cosmonauts would (1) assist in the checkout of the Super Booster and the moonship and (2) return to earth ahead of Apollo 11 in a Zond-type spaceship, escorting the unmanned Luna spaceship carrying several ounces of lunar soil. If the Apollo 11 astronauts met with misfortune, the propaganda value of the Communist victory would be hailed a million times over because it would have been accomplished at minimum cost and risk to their cosmonauts. This was the Soviet plan.

The sequence of events that occurred, as reported by reliable sources, is that on July 13, Luna 15 was launched by the Proton booster on a flight to the moon. As expected, this was the unmanned spacecraft that would attempt the landing and soil retrieval. Either before or immediately after the Luna 15 blast-off, the Soviets launched the unmanned Super Booster, but to their horror, the giant rocket blew up immediately after lift-off—years of intensive preparatory work destroyed in a ball of fire. The launch of the manned standard launch vehicle carrying the cosmonauts was cancelled and their hopes rested with the unmanned Luna 15 spacecraft.

On July 16, 1969, with the world watching, a glittering white Saturn V moon rocket carrying Apollo Astronauts Neil Armstrong, Edwin Aldrin, and Michael Collins lifted off gracefully amidst bright orange flames for the heavens; hours later Apollo 11 was heading silently towards the moon. On July 17, Luna 15 developed serious problems in orbit and Soviet controllers attempted to correct the problem by firing its rocket motors. This changed the spacecraft's orbital characteristics and permitted the Soviets to check out their guidance and control system. Hours later, the world heard Armstrong's voice during lunar descent: "Picking up some dust ... contact light ... engine stop ... Houston, Tranquility Base here. The Eagle has landed." After Armstrong stepped on the moon, Luna 15 attempted a soft landing, but again, due to anomalies in its guidance and control system and descent stage, Luna 15 crashed, dashing all hopes the Soviets had of returning lunar soil to earth before Apollo 11.

About fourteen months later, the unmanned Luna 16 spacecraft successfully returned several ounces of lunar soil to earth. With the success of this mission under their belt, the Soviets publicly criticized the United States for using the costly manned Saturn V moon rocket in its lunar expeditions when the task could be accomplished by "automatic means at less cost." Among other things, this was Communist double-talk to conceal their real intentions and work on the Super Booster, and for the most part, the U.S. news media bought the story; Americans were led to believe that the Russians withdrew from the moon race years earlier and were winding down their space program.

In 1969, I reported to the CIA that the first stage of the Super Booster can generate 10 million pounds thrust (compared to 7½ million pounds thrust for the Saturn V moon rocket), and when upper stages are added to the booster, it can inject about 330,000 pounds of useful payload into earth orbit (compared to 280,000 pounds for the Saturn V). When the Super Booster exploded, Academician Boris Petrov told his subordinates that the explosion set back their long-range plans by about three years and that "much redesign work is required now," but support for the program and their overall objectives were as firm as ever. (Some of this information was subsequently disclosed by *Aviation Week & Space Technology* magazine.)

The first stage of the Super Booster utilizes the advanced multichamber plug cluster engine concept that numerous U.S. engineers promoted for use in large rockets during the 1975–1980 time period. The multichamber plug cluster engine consists of about 24 rocket engine modules that are arranged in a circle and surround the upper part of a plug shaped similar to the plug of a bathtub. Plate 16A and 16B shows drawings of this type of booster concept from two KGB-approved Russian textbooks. Plate 16C shows a U.S. artist's conception of the Soviet Super Booster, based on the analysis of reliable information. The Soviet Super Booster concept also has separate steering rocket engines and special aerodynamic surfaces called shrouds that are not shown in the drawings. The shrouds guide the flow of air to the base of the vehicle to improve its performance.

The Soviets' problem with the Super Booster can be attributed to the advanced rocket engines it uses relative to those developed during the 1950s; these engines require special design provisions to ensure that the propellant in the propellant tanks is delivered smoothly to the rocket engines and distributed uniformly throughout the engines' combustion chambers. Because the Super Booster uses so many rocket engines, the Soviets encountered difficulties in meeting the necessary operating conditions for all the engines during their normal burn time. Glushko's design bureau is assisting the vehicle designers in this engine-vehicle integration problem, and there is no reason why the Super Booster will not be ready for another test flight in 1974.

With the Soviets quietly making fixes on the Super Booster after the 1969 explosion, I learned from their design engineers and space managers that when the vehicle becomes operational it will be used to support three major programs: (1) a manned lunar landing, (2) the establishment of enormous earth-orbiting space stations of a type never seen before, and (3) the establishment of scientific research bases on the moon. Several high-level Russians candidly told me that the Soviets will exploit the advantages of lunar base research because earth and orbiting research laboratories cannot simultaneously offer (1) a vacuum (no atmosphere), (2) low gravity, and (3) a stable research platform.

A prominent Soviet personality told me that the Soviets believe that they can achieve a scientific or technological breakthrough in atomic and laser physics through space and lunar base research. Accordingly, the Central Committee of the Communist Party approved plans to

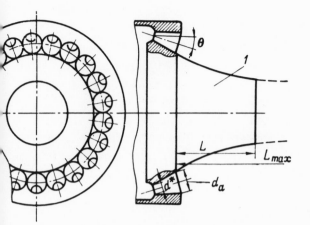

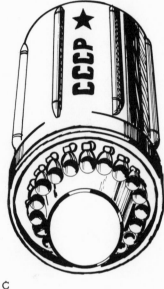

XVI-2-17. Сопловой аппарат в много-
сопловом варианте:
— максимальная длина центрального тела

A

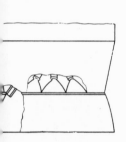

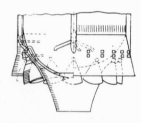

C

. 3.12. Многосопельная
гательная компоновка
внутренним расширением.

Рис. 3.13. Многосопельная
двигательная компоновка
с внешним расширением и
обрезанным центральным
телом.

B

PLATE 16 Drawings of Multichamber Plug Cluster Rocket: Sketches A and B are from Russian textbooks; Sketch C is artist's concept of the first stage of the Super Booster—based on reliable information.

establish both earth orbiting and lunar bases to support this effort. The impetus for the plan was provided by the late Academician Sergei Korolev, and in the middle 1960s, after his death, other academicians and his followers formed a coalition to ensure the success and completion of the program. When the Soviets land men on the moon, the event will be the prelude to an aggressive lunar base research program. The key to this program is the Super Booster.

In summary, the Soviets are known to be secretly developing a very advanced Super Booster that can place about 330,000 pounds of useful payload into earth orbit; this is about 50,000 pounds more than the lift capacity of the Saturn V moon rocket. More importantly, the Super Booster will be operational during the middle 1970s, when the manned U.S. space program is dormant, and will be used to support the establishment of lunar research bases and enormous earth orbiting space stations (discussed in later paragraphs).

SAS-20

To accommodate the launch of small scientific payloads in the hundreds of pounds range, the Soviets designed a small launcher called the Cosmos launch vehicle, which uses the RD-214 and RD-119 rocket engines in its first and second stages respectively. By American standards, this launch system is crude, but it does the job.

The Soviets have also made use of their existing ICBMs to inject payloads into orbit. (A discussion on Soviet ICBMs and the Soviet strategic threat occurs in later sections.)

The significant point that one should garner from the previous discussion is that Soviet launch vehicles have been designed to provide the Soviets with a well-planned manned and unmanned space program in the 1970s, and it is the objective of the Communist Party to expand this effort. While the Soviets developed versatile space launch vehicles that can be used for decades, most U.S. launch vehicles, particularly those relating to the manned space program, were designed with one objective in mind—to land men on the moon and bring them back home safely. In a sense, President Kennedy's directive in 1961 (" ... I believe that this nation should commit itself to achieve the goal, before this decade is out, of landing a man on the moon and returning him safely to earth") was both a blessing and a curse. It was a blessing because it united the nation's scientific and technological forces to achieve a single objective, but it was a curse because most U.S. planning was centered around the Apollo mission. Consequently, the Apollo program is now completed and the U.S. does not have the launch vehicles to pursue a viable space program for the 1970s. The technological base our nation built up through billions of dollars of expenditures is now spread among the ranks of unemployed engineers, or technical specialists who are employed in nontechnical professions. These and other subjects are discussed in later sections, but the main point is that during the 1950s and 1960s, the Soviets outplanned the United States in the development of space launch vehicles even though the United States got to the moon first; this will become apparent in the late 1970s, when the Soviets begin conducting space operations with salvos of space launchings.

SAS-21

Chapter 8

The Soviet Unmanned Space Program

In recent years, as Plate 17 indicates, the Soviets have been averaging about twice as many space launches as the United States. This has not been made clear to the man-in-the-street. Because of guidance, control, and general reliability problems, many of the probes launched by the Soviets in the late 1950s and early 1960s to the moon and planets were dismal failures. However, by the middle-late 1960s, the Soviets solved most of their problems and demonstrated an advanced capability to execute both civilian and military space missions with unmanned spaceships. Some of their more outstanding achievements with automated spaceships, which attracted the attention of U.S. intelligence analysts, are discussed below.

During October 1967 the Soviets launched two unmanned spaceships called Cosmos 186 and Cosmos 188. To the surprise of numerous Western observers, these spaceships rendezvoused and successfully docked using onboard guidance and control systems. Whereas Western newspapers reported that the Russian docking experiment was nothing to get excited about, because the manned Gemini spacecrafts docked in 1966, U.S. intelligence experts thought otherwise: the Soviet unmanned docking experiment was truly an outstanding technical achievement, and it demonstrated that the Soviets could conduct complex space operations such as linking up two

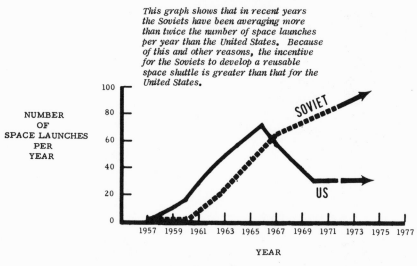

SPACE LAUNCHES
(GENERAL TREND)

This graph shows that in recent years the Soviets have been averaging more than twice the number of space launches per year than the United States. Because of this and other reasons, the incentive for the Soviets to develop a reusable space shuttle is greater than that for the United States.

NUMBER
OF
SPACE LAUNCHES
PER
YEAR

SOVIET

US

YEAR

PLATE 17   Number of Space Launches
SAS-22

or more spaceships and erecting orbital space stations, without the direct participation of man. Then, about a half year later in April 1968, the Soviets automatically linked the unmanned Cosmos 212 and Cosmos 213 spaceships, and they again announced that the spacecrafts were linked both electrically and mechanically. The significance of the mechanical link-up became apparent when I learned a year later in Paris that because of safety and cost considerations, Soviet space planners decided that for the 1970s their spaceships would have the capability to refuel in orbit, and mechanical link-ups would be similar to those of the Cosmos 186–188 and Cosmos 212–213 missions. For the 1970s this meant that propellant would be delivered into orbit using the low-cost standard launch vehicle and the Proton launch vehicle.

The automatic rendezvous and docking experiment also alerted the U.S. intelligence community that this "backward nation with the crude launch vehicle" possessed first-rate guidance and control systems and computers. Based on other intelligence information, the space experiments showed that the Soviets were not only developing the capability to erect orbiting space stations and perform refueling operations in space through unmanned systems, but they could ultimately rendezvous and intercept American orbiting satellites and other space systems.

It was mentioned earlier that fourteen months after the Luna 15 spaceship crashed on the lunar surface, Luna 16 returned samples of lunar soil to earth for examination by Soviet scientists. The unmanned Luna 16 mission caused some analysts in the U.S. to reassess the so-called computer gap between the East and West; some experts maintained that the Soviets lagged the United States in computer technology and applications, but Luna 16 and other Soviet accomplishments indicated that their computer effort was in many ways competitive to that in the United States.

Luna 17, launched in November 1970, showed that the Soviets had the capability to explore the moon using a roving vehicle called Lunokhod. Equipped with television cameras and looking very much like an old bathtub on wheels, the unmanned Lunokhod rover, guided by Soviet scientists in Moscow, traversed six and a half miles over the lunar surface for almost a year before it ceased to operate. Academicians Boris Petrov, Georgiy Ulanov, and Jouli Khodarev, who were mentioned earlier, played key roles in the development of Lunokhod and the soil retrieval Luna spacecraft.

In addition to the lunar missions, in the late 1960s and 1970s the Soviets used the Proton launch vehicle to send probes to Venus and Mars. The Russian probes not only weighed approximately five times as much as those launched by the United States, but the Soviets landed instrumented packages on both planets. The magnitude of the Soviet effort can be best understood in terms of the U.S. unmanned space program. The U.S. Viking Mission to Mars, our first attempt at an unmanned landing on a planet, is scheduled to be launched in 1975. Reliable sources have reported that the Soviets are already planning probes to Mercury, the planet closest to the Sun, and the outer planets of the solar system—Jupiter, Saturn, Uranus, Neptune, and Pluto. And this includes the Grand Tour Mission, which is an unmanned flyby of the outer planets in the late 1970s using a single space launch; this is possible because of the once-in-a-lifetime spatial arrangements of the outer planets (they are essentially grouped together). By comparison, the U.S. Grand Tour Mission was cancelled in January 1972 because of budgetary reasons; U.S. space officials hope to launch an unmanned flyby of the planets Jupiter and Saturn during the late 1970s, using modifications of existing hardware. This falls far short of Soviet plans, and history will certainly show that when the United States had the capability to explore the solar system our society yielded to the Soviets to perform this task.

In short, the Soviets will be very active during the 1970s with unmanned launches to the moon and planets to learn about the solar system, and in particular, to assist Soviet space research scientists. This activity, which also contributes to the Soviet technological base in areas such as the design of launch vehicles, space systems, computers, and control systems, can be best described in the words of Georgiy Zhivotovskiy (assistant to Vladimir Kirillin, chairman of the State Committee for Science and Technology): "If you pursue only that which you understand, soon you will have no science. You must be willing to gamble in science."

With the scientific and technological successes resulting from their earth orbital automatic rendezvous and docking experiments in 1967 and 1968, the Luna soil retrieval missions, the Lunokhod roving vehicle experiments, and the planetary explorations of Mars and Venus during the early 1970s the Soviets exhibited an aggressive unmanned space program that greatly exceeded the U.S. effort, and they clearly demonstrated the capability to conduct complex operations in space without the direct participation of man.

SAS-23

Chapter 9

The Soviet
Manned Space Program

By the late 1960s, U.S. intelligence analysts concluded that the Soviets outplanned the United States in the design of manned spacecraft systems; this has not been obvious to the American public because of the outstanding success of the Apollo lunar missions. The story begins in April 1961, when the Soviets launched Cosmonaut Yuri Gagarin into orbit in a spherical 10,400 pound spacecraft that the Russians called Vostok.

In February 1962, almost a year later, the United States launched Astronaut John Glenn into orbit in the 3,000 pound Mercury spacecraft. In October 1964, the Russians launched their first three man crew into space in a spherical 11,700 pound spacecraft called Voskhod. For all practical purposes the Voskhod spacecraft was a modified version of Vostok, and both were launched by the standard launch vehicle. About a half year later, in March 1965, the United States launched a two-man crew into space in a Gemini spacecraft that weighed about 7,100 pounds. In essence, the Vostok, Voskhod, Mercury, and Gemini spacecraft systems were test-beds for the development of more advanced manned spacecraft systems, and during 1961 through 1967, both nations conducted (a) orbital maneuvers in space, (b) extra-vehicular activities, (i.e., space walks), and (c) space rendezvous and docking experiments. The United States' effort was clearly aimed at developing the techniques that would ultimately support a manned lunar landing. Soviet intentions at this time were not clear but their effort was much broader, as if they intended to support both a manned lunar landing and an aggressive near-earth manned space program.

While U.S. intelligence analysts at the CIA headquarters in Langley, Virginia were arguing about the apparent direction of the Soviet space effort, the Soviets provided the answers to these and other questions themselves. In September 1968, the Soviets launched an unmanned Zond spacecraft around the moon and recovered it in the Indian Ocean. Then in October of the same year the Soviets launched a cosmonaut into orbit to check out a new 14,500 pound spaceship called Soyuz. American intelligence experts quickly noted the striking similarity between the Zond and Soyuz space systems, and on the basis of other information concluded that the Soyuz spacecraft would be used for near-earth space missions. This was significant because the Apollo program, managed by NASA, was designed specifically to support a manned lunar landing, and as is shown in later paragraphs, the Apollo spacecraft is not as versatile as Soyuz, nor can it support a viable manned near-earth space program.

To the surprise of numerous space specialists, the Soyuz spacecraft was made up of (1) a reentry module that can accommodate three cosmonauts, (2) a spacious spherical module called the orbital compartment that can be considered to be a mini-space laboratory because it was designed specifically to support the Soviet space research effort, and (3) a propulsion module for maneuvering in space. Compared to the Apollo spacecraft, the Russian ship affords the cosmonauts about one and a half times more working space.

Because of decisions made almost ten years earlier, the United States expended most of its space resources to explore the moon and has left near-earth orbital space to the Russians. In a flurry of classified memorandums and studies, U.S. space specialists concluded that the Apollo spacecraft could not be used for extensive earth-orbital space operations; it is too large due to the moon-related equipment and propellant carried aboard, and it does not have a mini-laboratory like Soyuz. Consequently, several government agencies recommended the development of another orbital spacecraft. In the midst of a Department of Defense and NASA controversy to determine the course of our nation's space effort, the Russians launched four cosmonauts into orbit in the Soyuz 4 and Soyuz 5 spaceships in January 1969, and as mentioned, they successfully rendezvoused, docked, and transferred two cosmonauts from the Soyuz 5 to the Soyuz 4 spacecraft. The

Soviets lauded this space mission as the creation of the world's first orbital space station. The docked Soyuz spaceships weighed about 29,000 pounds and had a combined interior working space that was about three times greater than that of the Apollo spacecraft. In addition to scientific space research equipment, the Soyuz orbital compartment contained a sofa and sleeping bags for the cosmonauts. The docking and crew transfer experiments indicated that the Soviets were also developing the capability to erect large orbital space stations in the future. Because the on-board guidance and control system for Soyuz was similar to that of the unmanned Cosmos 186–188 and Cosmos 212–213 spaceships that rendezvoused and docked automatically during 1967 and 1968, the Soviet manned and unmanned space programs complemented each other. It is worth mentioning that all of these spaceships were injected into orbit by the standard launch vehicle.

In summary, the Soyuz orbital compartment can be tailored to accommodate a variety of space missions, while the crew reentry module remains the same. Thus, Soyuz offers Soviet space managers considerable latitude in planning earth orbital space missions. The significant point about Soyuz is that it is both a manned maneuverable spacecraft and a mini-orbital space laboratory; the United States has no counterpart to Soyuz, even considering the Apollo spacecraft which was designed to support the manned lunar landings.

SAS-24

Because the Super Booster exploded in July 1969, Soviet space officials such as president of the Academy of Sciences Mstislav Keldysh and Academician Boris Petrov announced that the Soviets would continue to pursue the development of earth orbital space stations and explore the moon and planets by automatic means. As mentioned earlier, the international news media and American public bought this story, and the Soviets successfully conveyed the impression that they were involved in a low-keyed space effort.

By the fall of 1969, the Soviets were prepared to conduct further tests on the Soyuz spacecraft. On October 11, 12, and 13, they launched seven cosmonauts and three Soyuz spaceships (Soyuz 6, 7, and 8) into orbit. The cosmonauts performed orbital maneuvers, and the missions were highlighted by welding experiments performed in the orbital compartment of Soyuz 6. This demonstrated the Soviets' interest in perfecting space welding techniques to assist in the construction of permanent orbiting space stations. The mission also featured formation flights of the three spaceships. During these flights the cosmonauts experimented with methods of communicating with each other and used light sources that could not be monitored by normal electronic intelligence listening devices. They also conducted experiments to determine the visibility of objects at various distances from their spaceships, which among other things is the type of information used by military planners for designing equipment for photographing and inspecting hostile satellites.

As a result of the intensive Soyuz orbital flight-test program—Soyuz 2 through 8 were launched during October 1968-October 1969—the Soviets concluded that (1) orbital space stations could be constructed with the participation of cosmonauts, (2) crews of orbital stations could be replaced using the Soyuz spacecraft, and (3) cosmonauts very definitely had a role in space.

In July 1970, the Soviets continued their assault on near-earth space and launched a two-man crew in Soyuz 9. The cosmonauts stayed in earth orbit for 18 days, setting a long-duration space record, and obtained biomedical and psychological data applicable to long-duration space missions in both orbital spacecraft and space stations.

On April 19, 1971 the Soviets launched an orbital space station called Salyut with the Proton launch vehicle. A three man Soyuz 10 crew, launched four days later, attempted to visit the space station. Two members of the crew, Vladimir Shatalov and Aleksei Yeliseyev, participated in the earlier Soyuz 4 and Soyuz 5 rendezvous and docking experiment, and in the Soyuz 6, 7, and 8 formation flight missions. In addition to checking out the space station, the Soviets were obviously relying on the recommendations of these experienced space veterans to further improve the design of the Soyuz spaceship. However, the third cosmonaut, Rukavishnikov, became ill, and when the crew encountered technical problems they returned to earth.

In June 1971, Soyuz 11 was launched with another three-man crew. This time the cosmonauts visited the space station and conducted experiments for almost 24 days and set another long-duration space record. When the spaceship returned to earth, a leak in the hatch developed and the sudden decompression of air killed the three cosmonauts; a successful space mission was turned into a tragedy because the cosmonauts chose not to wear their spacesuits during reentry. Needless to say, the Soviet leadership was shocked when they learned of the deaths of the three cosmonauts. A subsequent investigation showed that the hatch to the Soyuz system had to be redesigned. The Soviets postponed further manned flights until the fixes on their inventory of Soyuz systems could be made. Technically speaking, the Soyuz program has already met Soviet expectations many times over, and their engineers are currently designing an advanced reusable orbital spacecraft system that will complement Soyuz in the late 1970s and replace it during the 1980s. (These and other advanced manned space systems that the Soviets are currently developing are discussed in the next section.)

In summary, U.S. space planners have left our nation with a dead-ended space program after the tremendous Apollo expeditions to the moon. We do not have the hardware to follow through after Apollo, particularly in near-earth space; our space program has been dismantled, hundreds of thousands of aerospace workers have been laid off, and the new frontier—space—has been left to the Russians. While the United States was expending most of its space-related resources on the manned exploration of the moon, the Soviets successfully pursued an aggressive near-earth manned space program and deployed (1) the versatile Soyuz spacecraft system that is both a maneuverable spaceship and "mini" orbital space station and (2) the Salyut space station. The Soviets gained valuable near-earth spaceflight experience, set records for long-duration space missions, learned how to maneuver and perform group flights—in 1969 they had seven cosmonauts and three spaceships in orbit simultaneously—and they studied problems relating to the development of advanced manned orbital spacecraft and space station systems. Unlike the Soyuz, the design of the U.S. Apollo spacecraft is not readily adaptable to exploiting near-earth space for scientific and military purposes. The significant point to note is that the Soviets will continue to conduct manned space missions throughout the 1970s using the low-cost standard launch vehicle and Soyuz spacecraft, and the Proton launch vehicle and Salyut space station. During this period the United States plans to launch only one Skylab space station (to be visited by three crews) and thanks to an agreement with the Russians, an Apollo crew will rendezvous and dock with a Soyuz spaceship in 1975. During the

1960s, the Soviets built a strong technological base in launch vehicle, spacecraft, and space station designs that will enable them to aggressively pursue the development of exceptionally advanced space systems in the future. The space systems that the Soviets are currently developing, which have greatly alarmed U.S. intelligence analysts and Air Force space planners, are discussed in the following sections.

SAS-25

Chapter 10

Soviet Operational Orbital Weapons Systems

The Strategic Arms Limitations agreement co-signed by President Nixon and General Secretary Brezhnev in Moscow on May 26, 1972 was another step in controlling the arms race by limiting the number of land and sea-based offensive and defensive missiles that can be deployed by the superpowers. Other major agreements reached by the two countries include the Moscow Treaty of 1963 banning nuclear weapons tests in the atmosphere, in outer space, and underwater; the Outer Space Treaty of January 27, 1967; and the Agreement on the Rescue of Astronauts and the Return of Objects Launched into Outer Space of April 22, 1968. Under the general conditions of the five year Strategic Arms Limitation agreement, subsequently ratified by the United States Senate by an 88 to 2 vote, and discussed in more detail in later sections, both the U.S. and U.S.S.R. are permitted to (a) make technological improvements of their existing intercontinental ballistic missile (ICBM) forces without increasing their numbers and (b) develop other earth-based weapons systems, such as long-range supersonic bombers, air superiority fighter aircraft, and advanced missile-firing submarines that can replace existing submarines. But what about the foreboding airless environment that shrouds the earth and extends for thousands of miles above its surface? None of the agreements with the Soviets precludes the use of near-earth space for military operations. For example, Article IV of the aforementioned Outer Space Treaty reads as follows:

> ... Parties to the Treaty undertake not to place in orbit around the Earth any objects carrying nuclear weapons or any other kind of weapons of mass destruction, install such weapons on celestial bodies, or station such weapons in outer space in any manner.

The treaty prohibits placing nuclear weapons or weapons of mass destruction into orbital space, but it does not preclude the use of orbital space for military operations, providing weapons of mass destruction are not employed. How do the Russians feel about military operations in space? One of the best authoritative Russian documents on the subject is Marshal of the Soviet Union V. D. Sokolovsky's 1968 publication, *Military Strategy*, which discusses Soviet ideology, military strategy, and the uses of orbital space. Sokolovsky states:

> The Soviet Union cannot disregard the fact that U.S. imperialists have subordinated space exploration to military aims and that they intend to use space to accomplish their aggressive projects—a sudden nuclear attack on the Soviet Union and the other socialist countries.

> In this regard Soviet military strategy takes into account the need for studying questions on the use of outer space and aerospace vehicles to strengthen the defense of the socialist countries. ... It would be a mistake to allow the imperialist camp to achieve superiority in this field. We must oppose the imperialists with more effective means and methods for the use of space for defense purposes.

What then are Soviet plans for the use of space and what is their capability? What space systems have they already deployed?

Both the United States and the Soviet Union have launched unmanned satellites to support military operations and assist strategic planners on earth. The list of unmanned military spacecraft systems includes (a) the so-called spy-in-the-sky satellites, which are electronic intelligence listening posts and unmanned spaceships that photograph air bases, naval harbors, orbiting spacecrafts, launch sites and silos and troop and ship disposition; (b) navigational satellites to assist military aircraft and ships at sea; (c) communications satellites used by the U.S. Department of Defense and U.S.S.R. Ministry of Defense to coordinate the activities and movements of their armed forces (d) warning satellites for detecting nuclear weapons in space and the launch of ICBMs on earth, and (e) other systems. These unmanned spacecraft systems generally support military activities on earth and do not interfere with the operations of another nation's satellites, aircraft, or ships. For this reason they are sometimes called passive space systems. Both the U.S. and U.S.S.R. aggressively support and conduct passive military operations in space.

An active or aggressive military space system is one that can interfere with activities on earth or the operation of another orbiting spaceship. Several U.S. intelligence analysts in key positions misled U.S. space planners during the middle and late 1960s when they incorrectly assessed Soviet space intentions in this area. For example, during the early 1960s, Premier Nikita Khrushchev and other prominent Soviet space officials boasted about building orbiting spaceships that could drop nuclear bombs on terrestrial targets. Experts within the U.S. intelligence community argued that this was idle talk, and maintained that the Soviets would not attempt to build spacecraft systems that could drop bombs from orbit because there were better ways of getting the job done (e.g., ICBMs and bombers can deliver a warhead to the target more accurately). The U.S. assessment was tainted with Western logic and methods of approach, and that was part of the problem. Just as the Soviets developed a launch vehicle in the 1950s that was as sturdy as a battleship and inefficient by U.S. standards, during the middle and late 1960s, the Soviets developed a fractional orbital bombardment system (FOBS) that could be launched into an orbit by an ICBM, pass over the United States, and at the discretion of the controllers in Moscow, release nuclear bombs and missiles to destroy our population, industrial centers, and our second strike B-52 jet bombers, which are stationed at various Strategic Air Command bases. While U.S. experts continue to downgrade the effectiveness of the FOBS weapon system, the Soviets believe otherwise and they invested millions of dollars to develop it.

Since 1968, the Soviets have been conducting significant orbital tests of an unmanned maneuverable spacecraft system that can ultimately be used to rendezvous with an American satellite, inspect it, and destroy it. During the middle 1960s, the United States was engaged in numerous classified analytical studies to see whether it was worth developing this capability. American planners decided against it, and one reason given was that an orbital satellite killer would not be cost-effective to develop. It is worth mentioning that the Soviet satellite killer uses storable propellants and a propulsion system that is simple, reliable, and durable. On the other hand, numerous studies in the United States on unmanned satellite killer systems emphasized the use of exotic and cryogenic propellants such as fluorine and hydrogen which required the design of very sophisticated and costly engines and associated space hardware. Again, the U.S. emphasis was clearly on performance, with cost being a secondary factor. However, when budgets were considered at a later date, most of the systems studied were neither cost-effective nor practical from a national position in relation to complementary space hardware. Additionally, some influential "experts" privately argued that it would not be necessary to develop an orbital satellite killer system because nuclear war would certainly break out if the Soviets destroyed one or more of our satellites, and under these circumstances the United States would rely on its ICBMs, Strategic Air Command bombers, and Polaris submarines. But this logic was clouded when the Pueblo intelligence ship was captured and the EC-121 reconnaissance aircraft was shot down by the North Koreans; it was subsequently concluded that the Soviets could destroy several of our satellites and the United States would not risk going to war. Furthermore, it was logically argued that depending on who is President of the United States, the Soviets could systematically destroy groups of our orbital satellites in secret and the United States would secretly tolerate the situation, just as Nikita Khrushchev secretly tolerated the U-2 flights over the Soviet Union until he could do something about them.

On the subject of satellite killers, the Soviets implicitly acknowledged that they indeed had the capability to destroy orbiting satellites when Foreign Minister Andrei Gromyko presented a draft treaty to the United Nations in October 1972. The Gromyko proposal essentially said that the Soviets would have the right to destroy a satellite if they believed that it was broadcasting illegal or erroneous information to its people. The United States rejected the proposal because it infringed on our traditional beliefs of free speech and uncensored information.

In summary, during the late 1960s and early 1970s, the Soviets conducted near-earth space operations necessary for developing a viable military space capability. The Soviet fractional orbital bombardment system (FOBS) and their satellite killer can be launched by a modified version of an existing ICBM. The significant point is that the Soviets have developed a capability to exploit near-earth space for military purposes without the direct participation of man, and they have accumulated orbital flight-test experience on military space systems that U.S. designers have studied on paper only. Though these active systems are currently operational, the real threat to Americans is the extension of this technology and experience to more advanced systems during the middle and late 1970s, when other space launch vehicles and space systems that the Soviets are secretly developing will become operational. This subject is discussed in the following sections.

SAS-26

Chapter 11

The Soviet
Space Shuttle Program

The most significant and advanced space development effort in the United States is the Space Shuttle Program. As mentioned earlier, the shuttle is a reusable space vehicle that should be operational in 1980 and will permit the United States to launch satellites into orbit for much less than it currently costs using the throwaway launch systems developed during the past twenty years. Serious work on the U.S. space shuttle effort began in 1969 when numerous corporations were contracted to study designs for both the space shuttle and the high chamber pressure rocket engines that would power the shuttle.

During 1969–1972 I had the unique opportunity to meet with the designers and planners of advanced Soviet launch vehicles. I not only learned about their design techniques, but after an extensive analysis I documented for the first time in classified circles the existence of a Russian space shuttle effort, including the names of the personnel and facilities that were supporting it, and details of the shuttle system that the Russians are designing. The paragraphs below discuss the Soviet space shuttle effort.

In 1965, planners from the Academy of Sciences, State Committee for Science and Technology, Ministry of Defense, Ministry of Aircraft Production, and others held a secret council meeting to discuss the future of the Soviet space effort, and specifically the development of reusable space launch vehicles. Prior to this meeting Soviet physicists achieved a major scientific breakthrough in laser physics and the engineering community was successful in its preliminary efforts to develop a lethal laser weapon system (the death ray). Accordingly, Soviet military planners devised elaborate, but preliminary, plans for using orbiting space stations and spacecraft armed with laser weapons to destroy ICBMs and ABMs launched by hostile nations from the earth below. Because of these and other reasons, the Central Committee, with the support of General Secretary Leonid Brezhnev, endorsed recommendations by Soviet planners to pursue an aggressive space program and develop a network of manned and unmanned orbiting space stations and spacecraft. Soviet planners logically argued that it would be expensive to support the planned space complex with food propellants, equipment, and other logistics supplies using existing throwaway launch vehicles developed during the 1950s and early 1960s (i.e., the standard launch vehicle, the Proton launch vehicle, and the Super Booster), so the decision was made to investigate the possibility of developing a reusable spaceship that could meet most of the Soviet civilian and military space logistics requirements through the 1980s. During this joint secret meeting Soviet planners agreed that the upper stage of the space shuttle would have to be powered by high chamber pressure oxygen-hydrogen rocket engines (to meet performance requirements), but they disagreed on the design characteristics of the first stage. The designers responsible for developing advanced aircraft and jet engines wanted to build an airplane-type first stage that was powered by very advanced jet engines, called scramjets. But the designers responsible for the development of missiles, space launch vehicles, and rocket engines, wanted to build an airplane-type first stage that was powered by high chamber pressure rocket engines. Because they could only agree on the type of propulsion system to be used in the upper stage, and this decision was straightforward since jet engines cannot operate in a vacuum (i.e., no air), Chief Designer of Spaceships Sergei P. Korolev and Chief Designer of Rocket Engines V. P. Glushko were given the go-ahead to study the design options for a rocket powered upper stage. Plans for developing the lower stage were postponed until both design groups (i.e., the jet engine and rocket engine designers) could substantiate their positions.

In 1966, Chief Designer of Spaceships Sergei P. Korolev died unexpectedly, and the rocket design bureaus under his direction were cannibalized and divided among less prominent chief designers. When another secret joint meeting was held to resolve the fate of the booster stage of the proposed Russian space shuttle, the arguments of the aircraft and jet engine designers prevailed; Korolev's death and the temporary disarray of the rocket design bureaus no doubt were factors in the decision to develop a space shuttle booster that would be powered by scramjets instead of rockets. Other points Soviet aircraft and military planners made were:

1. Advanced jet engines (scramjets) burn fuel more efficiently than rocket engines; thus, with scramjet engines powering the booster stage the space shuttle will be capable of delivering more payload into orbit per launch.

2. By developing a scramjet booster, the shuttle can be modified and used as an advanced bomber weapon system that can fly about twice as fast as current U.S. interceptor aircraft; it would be more logical to develop a system that could serve as both a space booster and a bomber. (Note: it would not be possible to develop an effective bomber from a rocket powered vehicle because rocket engines burn for only a few minutes, while advanced jet engines can run for hours.)

After an extensive evaluation of priorities, Keldysh's Academy of Sciences and Kirillin's State Committee for Science and Technology were ordered by the Council of Ministers to develop the space shuttle. Meanwhile, in the United States several corporations were independently conducting analytical studies on reusable space launch vehicles, but these studies were not coordinated on a national level.

By 1967 Russian design bureaus were very much involved in the development of a fully reusable rocket-powered airplane-type orbiter stage that could be injected into orbit and landed on an ordinary airfield for use over again. Soviet designers called this system the Rocketoplan. Preliminary Soviet estimates indicated that the Rocketoplan—the orbiter stage of the Soviet space shuttle—could become operational during the late 1970s. However, because the technology to develop the advanced jet engines for the lower stage—the scramjets—is very advanced, Soviet designers estimated that the booster could not become operational until the 1980s. Therefore, there would be an interim period where the Rocketoplan would have to be launched by mating it to an existing throwaway booster. This plan was acceptable, because the Soviets planned to conduct flight-tests with the Rocketoplan before finalizing its design, and using existing hardware was the low-cost way to do it. In the United States the Air Force began to develop a high chamber pressure oxygen-hydrogen rocket engine, but no organization was developing the reusable launch vehicle that would be powered by these rocket engines.

By 1968 the secret Soviet space shuttle program was going full steam ahead. It was planned and supported by leading Communist Party officials, and a space shuttle commission was created to advise, manage, and coordinate the development of both the Rocketoplan and the reusable booster stage. The commission decreed that the Moscow Aviation Institute and V. P. Glushko's rocket engine design bureau were responsible for the development of the Rocketoplan, and the Central Aerohydrodynamics Institute (TsAGI) would manage the development of the advanced scramjet booster. Russian research teams and institutes from the Academy of Sciences and State Committee for Science and Technology were aggressively researching and developing components for the scramjet engines that would power the reusable booster; project managers were also assigned to follow the Rocketoplan program to ensure that the Rocketoplan was being designed so that it could first be mated to an existing throwaway rocket booster, and eventually the reusable scramjet booster—the lower stage of the Russian space shuttle.

During this period Soviet planners from the Ministry of Defense finalized their mission studies which showed that the Russian space shuttle could be justified on both an economic and military basis: the reusable scramjet booster could be modified for use as an advanced bomber weapon system, and the Rocketoplan could eventually be designed with a capability to destroy hostile satellites, spacecrafts, and space stations, and when armed with bombs, missiles, or laser weapons it would be capable of destroying terrestrial targets. Additionally, the shuttle could deliver military hardware into orbit at low cost.

In early 1969 experts in the United States began promoting the development of a reusable space shuttle, but concluded that it was still necessary to upgrade the state-of-the-art in propulsion and material technology before the shuttle could be built. It was argued that if a commitment were made to develop a reusable launch vehicle, the system's rocket engines would take the longest to develop. This was one reason why the United States Air Force had the foresight to begin development of a high chamber pressure oxygen-hydrogen rocket engine in 1967 for possible use in a future reusable launch system. But in 1969 the Air Force was still developing the engine, and no government agency was developing the launch vehicle to go with it. And a reusable launch vehicle was not being developed because the United States had no firm space goal after the Apollo program.

The first major step in this direction was initiated by President

Nixon as soon as he took office in 1969. In a memorandum to the secretary of defense, the acting administrator for NASA, and other officials, President Nixon requested recommendations on the direction that the U.S. space program should take in the post-Apollo era. While the report to the President was being prepared, insiders were told that one of the recommendations would include the development of a reusable space launch vehicle—the space shuttle. Accordingly, both NASA and the Department of Defense initiated serious space shuttle preliminary design studies.

During the spring of 1969 I learned in Paris that the Soviets had already flight-tested an unpowered prototype of their Rocketoplan; it was dropped from a military aircraft flying at high altitude and its aerodynamic performance characteristics were monitored up to the time it landed at an airfield. These tests were conducted without rocket engines, and in some ways were similar to the classified flight-tests conducted by the U.S. Air Force on small airplane-type reusable spacecraft at Edwards Air Force Base.

When the Soviets learned that the United States was embarked on a space shuttle program of its own, the KGB went to work. What configurations and payload values were U.S. designers considering? What were their specifications? More importantly, what role would the military play in its design and use? In what was one of the most organized yet subtle intelligence efforts in modern times, the Soviets sent some of their best scientists and space shuttle experts to international conferences throughout the world, and even to the United States, where most aerospace companies were openly presenting reports on their preliminary space shuttle designs at technical meetings. Within months the Soviets learned that the U.S. had a serious problem: two U.S. government organizations were conducting operations in space; the National Aeronautics and Space Administration and the United States Air Force. Most of NASA's space program was civilian-oriented, aimed at exploiting the peaceful uses of outer space. The USAF's interest was naturally military-oriented. In the past both government organizations developed launch vehicles and space hardware to serve their purposes. During the Sputnik era the public, and particularly Congress, tolerated redundant efforts by both organizations. NASA developed the Saturn 1 and Saturn 1B launch vehicles that could deliver approximately 22,000 and 37,000 pounds payload into earth orbit respectively; the USAF developed the Titan III-C launch vehicle that could deliver approximately 25,000 pounds payload into orbit. It was obvious, however, that each organization could not be funded to develop its own reusable launch vehicle; the cost would be prohibitive.

Because the Soviets do not have separate military and civilian space agencies (i.e., the Academy of Sciences manages the Soviet space program for the military) the Soviets developed the minimum number of launch vehicles and space hardware, and all of their launches are performed by the Strategic Rocket Forces of Grechko's Ministry of Defense. Therefore, comparatively speaking, the United States operated under a disadvantage: whereas the Russian space program is managed by the Academy of Sciences for the military in accordance with the guidelines and directives of the Communist Party, the United States, to prove to the world that we intend to explore the peaceful uses of outer space ("for the benefit of mankind") created a civilian space agency (NASA) to manage its space program. Furthermore, by the late 1960s, NASA programs had to be justified before Congress on a scientific, technological, and economic basis. Due to an inept government public relations program, a sophisticated KGB propaganda program, and a disinterested and slanted press, the American public has been brainwashed into thinking that it is immoral for the United States, and above all, for NASA to even consider using space for military purposes, yet it is accepted that this is the Soviets' prerogative.

During the late summer of 1969, many American aerospace engineers with secret clearances knew that both the U.S. and U.S.S.R. were involved in extensive and detailed studies on reusable launch vehicles, but few intelligence analysts knew that the Soviets were more involved in this area than the United States. The Soviets were not just conducting paper studies; they were building hardware. As was the case with the United States, the Soviets have considered a Rocketoplan that can temporarily deorbit, fly over hostile territory like a glider, drop bombs or launch missiles, and then return to a safe orbit. The possibilities are limitless, and as mentioned earlier, the Rocketoplan could also be used to destroy hostile satellites, spaceships, and space stations. To the delight of numerous Soviet officials, the decision was made within the Executive Office to permit NASA to develop the U.S. space shuttle. This meant that the U.S. space shuttle would be designed by a civilian agency with possible guidance from the military, instead of vice versa. Thus, the desirable military performance characteristics of the U.S. shuttle would be compromised to serve civilian objectives, whereas in the Soviet Union, the space shuttle was being designed with military objectives in mind.

During this period, I began putting together a story on the organization of the Russian space shuttle program, and it appeared that some of the key Russian personalities involved in the effort would attend the XX International Astronautical Federation Congress in Mar del Plata, Argentina. I sent a memorandum to my CIA contact in August 1969 because I was aware of the KGB involvement at these meetings, and advised him that I would be covering the Argentine meeting with the objective of learning more about the Russian shuttle effort. Appendix I contains excerpts from this memorandum and shows some of the questions that I hoped to have answered at the meeting.

In Mar del Plata, the Soviet intelligence effort was aimed at learning more about U.S. space shuttle plans from U.S. scientists. It was a give and take confrontation, and both sides had something to learn. Some of the prominent personalities I held lengthy discussions with included Oleg Belotserkovskiy (head of the Moscow Physical Technical Institute), Jouli Khodarev (deputy director of the Institute of Space Research), Vladimir Sychev (deputy director of the Central Aerohydrodynamics Institute, TsAGI), Gennadi Dimentiev (one of the secret heads of the Soviet space effort, and affiliated with the Moscow Aviation Institute), Georgiy Zhivotovskiy (KGB and currently assistant to the deputy premier of the Soviet Union, Vladimir Kirillin), Youri Zonov (KGB and chief of Scientific Information at the Institute of Space Research), and Igor Prissevok (KGB watchdog and frequent companion of Soviet cosmonauts when traveling abroad (See Plate 18). In what was a most productive week, I confirmed that the Russians were developing a space shuttle that would use rocket engines in the upper stages, and ultimately scramjet engines in the booster stage. Both Dimentiev and Sychev said that the United States was making a grave mistake by pursuing the development of a reusable launch vehicle that was powered by rocket engines in both stages, and they disapproved of the preliminary designs of the U.S. shuttle. The Russians made the following points:

(1) The United States should not engage in a space shuttle development program if the payload capability of the system is only 25,000 pounds (NASA's space shuttle specifications during 1969). I was bluntly told that the Soviets would not consider developing a space shuttle unless considerably more than 50,000 pounds payload could be delivered into orbit per launch.

(2) The propellant weight-structural weight characteristics of the U.S. shuttle's upper stage invited costly degradation in payload capability when the system was finally built because of unforeseen

PLATE 18   With Russians in Mar del Plata, Argentina

design changes. (Note: in rocket design language, the numerical value of a stage's propellant weight, divided by the sum of its propellant weight and structural weight, is called propellant mass fraction. This number is always less than 1.0, and the more efficient a stage is designed, the closer this number approaches 1.0, and the higher the payload capability of the system. The value of this fraction in the upper stage shuttle configurations studied by U.S. companies in 1969 varied between 0.6 and 0.7, depending on the system selected and the amount of thermal protection used.) With KGB specialist Igor Prissevok translating, Gennadi Dimentiev told me: "If the United States can't design a reusable upper stage with a propellant mass fraction of 0.8 or greater, then the United States should forget the whole idea."

In short, the Russian was saying that it is possible to invest billions of dollars in a space shuttle that might not work. One of the reasons for this was that the United States was designing small shuttle stages, whereas the Russians characteristically build large systems. The Russians maintained that NASA's design specifications allowed no room for error, and emphasis was still being placed on smallness and sophistication instead of simplicity and overall mission objectives.

By 1970 the Soviet space shuttle effort was at least three years ahead of the American effort, and the gap was growing because the United States was embarked on a shuttle program that could not be supported by the available funds, which meant that sometime in the future the program would have to be redirected. My second annual secret report was completed in mid-1970, and I made the following points:

(1) The Russian space shuttle is being designed to deliver between 75,000 and 100,000 pounds payload into orbit, while NASA studies considered shuttle payload values of 25,000 to 50,000 pounds. Because of long-range military and cost considerations, it is in our national interest to significantly increase the payload capability of our shuttle. Just as the Russians outplanned us in the 1950s, when they built a standard launch vehicle that could deliver tons of payload into orbit, while we were designing systems that could deliver pounds of payload into orbit, the Russians were now designing a space shuttle system that could deliver significantly more payload into orbit per launch than the U.S. shuttle system; they were again thinking well into the future and in terms of large and bulky military space systems that could be placed into orbit during the 1980s by their Rocketoplan.

(2) Because of cost and design considerations, the Russians planned to use a throwaway booster at first, while NASA studies considered the

design of a costly manned rocket-propelled reusable booster that could land at an airfield.

(3) The Russian space shuttle was military-oriented, while NASA (civilian) and the DOD (military) were involved in a major controversy over the design specifications of the shuttle.

During this period, Max Faget, a chief design engineer at Houston's NASA Manned Spacecraft Center who conceived the Mercury spacecraft and contributed to the design characteristics of other NASA spacecraft systems, introduced a new reusable space shuttle concept in contrast to the systems already being studied by NASA headquarters in Washington. Faget's system could deliver 25,000 pounds payload into orbit and reenter the earth's atmosphere at very high angles of attack, resembling an airplane descending with its tail drooping significantly. The system was attractive because of aerodynamic and heat transfer considerations, but the Department of Defense did not like the concept because it offered low cross-range capability, which meant that the orbiter stage could land only at a limited number of airfields, located mostly in the direction of the stage's reentry line of sight. For comparison, the Soviet Rocketoplan (See Plate 19) is being designed with a high cross-range capability so that it can fly over more territory during reentry. This is important to Grechko's Ministry of Defense because the Rocketoplan can be deorbited immediately upon command from Moscow and landed at one of many possible landing sites; the Rocketoplan is being designed with a quick reaction time capability, a characteristic trait of most military weapons systems. To underscore the Soviet emphasis on high cross-range capability, Vladimir Sychev, deputy director of the Central Aerohydrodynamics Institute—the institute responsible for developing the booster stage of the Russian shuttle—approached me at a meeting in October 1970 and said that he could not believe that the United States was seriously considering the Faget concept for its shuttle. When I asked Sychev to elaborate, the deputy director said that it was not practical to spend billions of dollars to develop a shuttle system that could not fly over more territory. Being one of the heads of the facility in the Soviet Union that manages, develops, and tests major Soviet aircraft weapons systems, Sychev's comments were not those of an ordinary scientist or engineer. To accentuate Soviet interest on the military applications of the shuttle, during a flight between Buenos Aires and Mar del Plata a Soviet design engineer from the Moscow Aviation Institute bluntly asked me whether the first stage of the American space shuttle had the capability to fly over the Soviet Union after being launched from Cape Kennedy.

PLATE 19 Artist's Concept of Russian Rocketoplan—based on reliable information

In 1971, because of budgetary, political, military, and other reasons, NASA redirected its space shuttle program. Unknown to most Americans, the redirected effort and major conceptual design changes emulated the Russian shuttle program during the 1968–1969 time period as documented in my secret report and private memorandums to Pratt & Whitney Aircraft management and the CIA in 1970:

(1) The shuttle's payload specification increased from 50,000 to 65,000 pounds (note: Russian shuttle is being designed to deliver between 75,000 and 100,000 pounds payload).

(2) The second stage conceptual design was changed so that its propellant weight was high relative to its structural weight (note: this relates to the propellant mass fraction comments by Gennadi Dimentiev—see page 133).

(3) A manned airplane-type booster powered by rocket engines would not be built (note: this relates to the comments by Gennadi Dimentiev and Vladimir Sychev—see page 131).

(4) The orbiter stage would be designed with a high cross-range capability, which meant that it could fly over more territory during reentry to accommodate the Department of Defense's military requirements (note: this relates to comments by Vladimir Sychev—see page 134).

Because of budgetary problems and congressional pressure, the U.S. space shuttle effort was also separated from the context of a total national space program. The programs recommended by President Nixon's Space Task Group in 1969, such as the development of space stations and reusable orbital spacecraft that would operate entirely in space, were deferred. The tumultuous redirection of the U.S. space shuttle program, coupled with the decision to defer development of other vital programs, were regarded by numerous aerospace engineers and managers as a bad planning decision by NASA; it was argued that the space shuttle was needed to support an aggressive space program in the 1980s, and this included space stations and orbital spacecraft. With the deferred development of these space systems, the logical question asked in private circles was: why build the shuttle?

By comparison, the Russians were planning their space shuttle effort in the context of a total national space program; they decided in the middle 1960s to develop a network of earth orbiting space stations and reusable orbital spacecraft (discussed in later paragraphs) in addition to their shuttle. Plate 17 showed that in recent years the Soviets performed about twice as many space launches as the United States. Because the Soviet launch vehicle traffic in the future will increase even more, the development of a reusable space shuttle for lowering launch vehicle costs is a sound investment; the incentive for the Soviets to develop the system is greater than that of the United States, and the principal reason for this is that the Soviets have definite plans for using space. On the other hand, the United States was again using the costlier piecemeal approach; the managers who knew about the Soviet effort argued: "When the public realizes what the Russians are up to, they'll support anything we recommend." This logic, which still prevails in some circles, has been criticized by concerned insiders as being costly, shortsighted, inconsiderate to the American taxpayer, and dangerous to our national security.

When Dr. Wernher von Braun left NASA in 1972, and some of NASA's planners were reshuffled within the U.S. government, it signalled an era where logic and planning in our space program is practically nonexistent. Reliable sources have subsequently substantiated my earlier assessment of the Soviet space shuttle effort, and there is every reason to believe that the Rocketoplan—the orbiter stage of the Russian shuttle—could become operational by 1976. Additionally, a highly regarded Eastern bloc source reported that the Soviets were encountering difficulties in the development of the scramjet propulsion system that would be used in the reusable booster stage, and a tentative decision was made by Soviet planners to use both jet engines (scramjets) and rockets in the booster. The reusable booster—mated to the Rocketoplan—is expected to become operational in the early 1980s. (See Plates 20 and 21.)

PLATE 20 Advanced Russian Reusable Launch Vehicle Concepts *LEFT*—Russian drawing of an advanced booster (photograph taken at the Soviet Space Exhibit in Belgrade, Yugoslavia in 1967). The upper stage shows part of a winged vehicle. *RIGHT*—U.S. artist's concept of similar launch vehicle with Rocketoplan (second stage). This is one of several space shuttle designs considered by the Soviets during the middle-late 1960s.

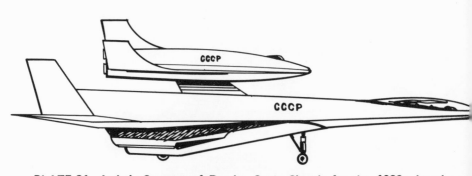

PLATE 21 Artist's Concept of Russian Space Shuttle for the 1980s—based on reliable information. The Rocketoplan rests atop the scramjet booster and the shuttle takes off like an airplane.

By the end of 1972 the United States' space shuttle program was well under way, and North American Rockwell Corporation, selected by NASA to design the shuttle's orbiter stage, was given the go-ahead to proceed with the development of the system. Rocketdyne was selected in 1971 to develop the high chamber pressure rocket engines that would power the orbiter stage. Even with the 1971 redirection of the shuttle program, the Russian Rocketoplan will still deliver considerably more payload into orbit per launch than the American shuttle. Additionally, the larger payload compartment of the Rocketoplan is designed to accommodate bulky military payload systems. For the booster stage, the United States is committed to develop large solid rocket motors that can be recovered in the ocean after launch for use over again. Inside observers following the shuttle efforts of both nations have privately stated that from the very beginning the Russians have been pursuing a well-planned methodical approach towards developing a space shuttle, while the U.S. effort has been greatly tied in with politics. In fact, NASA cannot afford to redirect the shuttle effort again because of congressional resistance, even though private corporate studies in December 1972 have shown that NASA might be developing the wrong space shuttle system relative to our defined space objectives. Some have argued that it does not make sense to spend billions to develop a reusable booster that must be recovered in the ocean in the 1980s, when the Russians will be operating their space shuttle like an airline company, with both stages ultimately being able to take off and land on a runway; if this is to be the end result of our long-range multi-billion dollar space shuttle effort then we are in deep trouble, because it is no secret that the Department of Defense will be the primary user of NASA's space shuttle when it becomes operational. And if NASA plans to replace the solid rocket boosters that must be recovered in the ocean with an airplane-type reusable booster, then Congress should be told about it because the public is not in the mood for surprises, and the reusable upper stage of the shuttle must be designed with this in mind.

An analysis of the situation shows that the United States is committed on this course because the twenty year space shuttle program (about seven years of development and thirteen years of operation) is being planned during a period of political controversy, when a mood of anti-space and anti-defense spending is sweeping the nation, and when NASA's budget is at an all time low.

NASA needs the space shuttle program to stay alive—without the shuttle program there would be massive layoffs within NASA and the aerospace industry—and it is the general feeling within the engineering community that the shuttle is really a make-work program; the idea of planning a long-range space program at the lowest possible cost is of secondary importance. And the public cannot judge the course of the U.S. space effort because pertinent information about the Soviet space program has been withheld from them. Even though the United States landed men on the moon first, the truth of the matter is that the Soviets have outplanned the United States in their effort to exploit the scientific and military uses of space, and this will become obvious in the late 1970s when the Soviets systematically unveil their new space hardware and weapons.

Tables IV and V summarize the American and Russian space shuttle programs. The significance of the Russian shuttle and other weapons systems are analyzed in later sections.

The academic-minded reader might find it interesting to read several raw intelligence reports that I prepared for the CIA that relate to the Russian space shuttle effort. (See Appendix II.) These reports are termed raw because they were based on my contacts with the Soviets in Europe and South America, and were prepared independently of existing intelligence information. Finished intelligence reports are written by analyzing raw intelligence reports and ferreting out inconsistencies, erroneous information, and speculative comments. The raw intelligence in Appendix II appears as it would to U.S. government intelligence analysts, with the exception that certain passages have been deleted to protect sources of information from possible recrimination. Information on the Russian space shuttle has been compartmentalized within the U.S. intelligence establishment because certain empire-

SOVIET UNION

UNITED STATES		SOVIET UNION
	1966	U.S.S.R. Academy of Sciences and the State Comm[ittee] for Science and Technology of the U.S.S.R. Council [of] Ministers are given go-ahead to develop Space Shut[tle]
	1967	Russian design bureaus are given go-ahead to deve[lop] fully reusable rocket-powered, airplane-type orbit[er] stage, which they call the "Rocketoplan"; decision [is] made that first stage of Space Shuttle will use exist[ing] "throw-away" rocket booster. Operational target d[ate] is late 1970's.
	1968	In parallel effort, Russian research teams are give[n] go-ahead to research and develop scramjets (i.e., very advanced jet engines) for use in airplane-type reusable first stage (booster). This advanced first stage eventually will replace the existing "throw-aw[ay]" rocket booster in the Russian shuttle. Operational target date is 1980's. Russian Space Shuttle Commission decrees that Mo[scow] Aviation Institute and rocket design bureaus are re[s]ponsible for development of "first generation" Spac[e] Shuttle (i.e., the "Rocketoplan" mated to "throw-a[way]" booster). It is also legislated that Central Aerohy[dro]dynamics Institute is responsible for development [of] airplane-type reusable booster that will replace "[throw] away" booster.
NASA embarks on serious Space Shuttle definition studies; most studies consider fully reusable airplane-type systems powered by rocket engines.	1969	Russians send Space Shuttle experts to United Stat[es] international conferences to learn about U.S. shutt[le] effort. Russians privately criticize U.S. Space Sh[uttle] plans: say U.S. payload specifications of 10,000-5[0,000] pounds are too low; say U.S. second stage propell[ant] weight is too low relative to structural weight; say [U.S.] should not use rocket engines to power airplane-ty[pe] reusable booster; say U.S. shuttle design is not ve[ry] enough.
NASA initiates detailed design studies of a fully reusable airplane-type two stage shuttle powered by rocket engines; proposed payload capability per launch is 50,000 pounds; some NASA centers propose Space Shuttle designs that can deliver 10,000 pounds payload into orbit. Secret report (PWA-FR-3760A) documents Russian Space Shuttle effort for first time and is distributed to Executive Office of President and NASA in November 1970.	1970	Shuttle program on schedule and planned in the co[ntext] of an aggressive manned and unmanned total natio[nal] space program per guidelines from the Communis[t] Party of the Soviet Union. Shuttle payload capability in the 75,000-100,000 p[ound] range to accommodate aggressive space program [in] coming decades.
Major and tumultous redirection of U.S. Space Shuttle program because of budgetary, political, and technical reasons. Changes emulate Russian position and advice in 1968-1969: . Payload capability of shuttle increased to 65,000 pounds per launch . Second stage of shuttle redesigned so propellant weight is high relative to structural weight . Airplane-type booster powered by rocket engines eliminated . Orbiter stage landing characteristics adjusted to satisfy military requirements Space Shuttle program separated from context of total national space program because of budgetary and planning considerations; programs recommended by President Nixon's Space Task Group in 1969, such as reusable orbital spacecraft and space stations, are either deferred or eliminated.	1971	Shuttle program on schedule.
North American Rockwell Corporation selected to develop orbiter stage of Space Shuttle; decision made to use solid-propellant booster that can be recovered in ocean for use over again. NASA slows down work on Space Shuttle because of budgetary reasons.	1972	Soviets tighten secrecy on space shuttle effort; fe[w] scientists with knowledge of space shuttle progra[m] permitted to travel abroad.

TABLE IV Space Shuttle Chronology
SAS-27

SPACE SHUTTLE COMPARISON (PRESENT PLANS)

	US	Title Heading	SOVIETS	
	First Generation Shuttle*		First Generation Shuttle	Second Generation Shuttle
Payload to Orbit per launch	65,000 pounds		75,000-100,000 pounds	75,000-100,000 pounds
Number of Stages	2		2	2
Description of Upper Stage	Reusable airplane-type orbiter stage powered by rocket engines		Reusable airplane-type orbiter stage powered by rocket engines (The "Rocketoplan")	The "Rocketoplan"
Description of Booster Stage	Reusable vertical lift-off solid rocket-type; ocean recovery		Throwaway vertical lift-off rocket-type	Reusable horizontal take-off airplane-type first stage; airstrip landing; powered by Scramjets (advanced jet engines)**
Operational Date	1980's		Late 1970's	1980's

US:

Effectiveness of program is very dependent on the development of reusable orbit-to-orbit spaceships and orbital space stations, which this country is not committed to develop in near future; U.S. continues to use piecemeal approach to conquering space; on long-term basis this approach is costlier, it places a heavy burden on taxpayer, and it is dangerous to national security.

*Second generation Space Shuttle not defined by NASA

SOVIETS:

Program integrated with planned development of a network of orbital space stations and reusable orbit-to-orbit spaceships to exploit the scientific and military uses of space; developed within the context of a total national space and defense program; is conservative, methodical, well-planned, and emphasizes the use of existing and proven hardware; first generation Space Shuttle designed and coordinated with plans for second generation shuttle.

** With slight modification, can be converted into a long-range bomber weapon system that can travel at speeds greater than seven times the speed of sound (more than twice as fast as current operational interceptor aircraft)

TABLE V Space Shuttle Comparison (Present Plans)
SAS-28

builders, for reasons that need not be mentioned here, have been withholding pertinent data from their colleagues and the users of the intelligence. Due to a gross lack of coordination between the CIA and the military intelligence services, the United States is not using its intelligence resources efficiently. I have identified the key personalities and facilities involved in the Russian space shuttle program in Appendix III. Sovietologists in the West, who have the time and resources, are encouraged to overtly pursue the shuttle-related activities of these men and the work at the facilities listed. Significant information or analysis should be reported to their respective governments, and if possible the information should be reported publicly through the news media. Public disclosures would force U.S. government intelligence analysts to privately justify to their superiors those assessments that differ from the public version; this is something that is currently not required of analysts within some sectors of the U.S. intelligence community.

Chapter 12

The Soviet Reusable Orbit-to-Orbit Shuttle Program

In 1969, planners within the Department of Defense were informed that the Soviets were embarked on an aggressive reusable orbital spacecraft development program in addition to their space shuttle effort. Air Force officials immediately grasped the significance of the Russian project: classified studies, including my own, showed that a space shuttle would not be effective unless accompanied by a separate reusable orbital spacecraft system. The logic went something like this: when the second stage of the shuttle (i.e., the North American Rockwell orbiter-stage or the Russian Rocketoplan) is injected into an earth orbit, its propellant tanks are nearly empty (i.e., most of the propellant is used just to get into a low earth orbit). Therefore, the shuttle's range in near-earth space is limited and it cannot perform certain missions that are critical to the military. For example, the space shuttle will operate in a space environment that is generally several hundred miles above the earth, but some of the Air Force secret satellites must be injected into a circular orbit that is about 22,300 miles above the earth; this is the so-called synchronous or 24 hour orbit that permits a satellite to essentially hover over a fixed point on earth. To perform this and other high energy missions, U.S. space planners concluded that a fully fueled small spacecraft that can be carried into orbit in the payload compartment of the space shuttle would be needed. Then, this vehicle—called the Space Tug by NASA planners and the Orbit-to-Orbit Shuttle by Air Force planners—could carry out those space missions that would be beyond the capability of the space shuttle. In layman's language, the principal behind the space shuttle and the orbit-to-orbit shuttle can be described as follows: Arlington House Publishers wishes to distribute this book on the West Coast. The publishers therefore load a tractor trailer with gasoline, oil, one fully fueled Volkswagen, and fill the remaining trailer space with copies of this book. The trailer is driven from New York to Los Angeles and book deliveries are made with the Volkswagen. The Volkswagen would get better gas mileage than the tractor trailer, and the tractor trailer could be used as a refueling depot when the Volkswagen gets low on gas. This is essentially the principal behind the space shuttle (tractor trailer) and orbit-to-orbit shuttle (Volkswagen) system.

In 1969, U.S. space planners recognized that both the space shuttle and the orbit-to-orbit shuttle had to be built; without the orbit-to-orbit shuttle the space shuttle would be as good as a tractor trailer out of gas on the Los Angeles freeway. By 1970, a detailed evaluation of the Russian orbit-to-orbit shuttle program showed that the Russians were pursuing an entirely different design approach than the United States. In earlier chapters it was mentioned that Soviet designers place emphasis on the storability of propellants and extensive refueling operations in space. The Soviets, therefore, elected to use storable propellants in their orbit-to-orbit shuttle. By comparison, U.S. Air Force design teams initiated detailed studies on a sophisticated orbit-to-orbit shuttle system that used supercold oxygen-hydrogen (cryogenic) propellants.

Just as the U.S. Air Force Aeronautical Systems Division routinely requires intelligence information on Russian fighter aircraft to determine the design specifications of future American fighter aircraft, the U.S. Air Force Space and Missile Systems Organization requires information on the Russian orbit-to-orbit shuttle to assist them in planning the U.S. military space effort. By the end of 1970 I put together the following story on the Russian orbit-to-orbit shuttle, based on reliable Eastern bloc sources and personal conversations with the designers of Russian spacecraft. (Some of the information in the following paragraphs was summarized in my second annual secret report on Russia's space program, but because of political reasons the U. S. Air Force Space and Missile Systems Organization, which is heavily involved in orbit-to-orbit shuttle design studies, was excluded from seeing the analysis, even after it officially requested a copy of the report from Pratt & Whitney Aircraft.)

The Russian orbit-to-orbit shuttle program is being coordinated from the highest levels and is planned in the context of a total national space program. It will become operational in the late 1970s and eventually replace the Soyuz spacecraft. Whereas the Soyuz is used once and then discarded, the Soviet orbit-to-orbit shuttle is being designed to be used again and again in orbital space; cosmonauts and fuel will be delivered to and from the orbital spacecraft by the space shuttle. The design of its command module is based on the Soyuz program flight experience and recommendations made by Cosmonauts Shatalov and Yeliseyev, who flew joint missions in Soyuz 4, Soyuz 8, and Soyuz 10. Its propulsion module will use reusable and durable rocket engines designed by the V. P. Glushko design bureau; Chief Designer Aleksei Isayev was involved in the design of this rocket engine system before his death. The Russian orbit-to-orbit shuttle will use strap-on reusable cylindrical propellant tanks that can be attached to the core of the spacecraft; by varying the number of propellant tanks that surround the core, the total weight of the Russian spacecraft can be varied from about 20,000 to 100,000 pounds. (See Plate 22.) Soviet space planners have concluded that an orbital space system must employ modular spacecraft construction and have the capability of being refueled in orbit. By using strap-on propellant tanks, the Russians can tailor their spacecraft to accommodate specific missions. For example, more propellant tanks would be used for high energy deep space missions. This can be done by storing propellant tanks in a propellant depot near orbital space stations and assembling the spacecraft in orbit by automatic means. These operations are made easier by using propellants that can be stored and transferred easily. To duplicate the Soviet plan, the United States, which has promoted the use of cryogenic propellants for orbital spacecraft, must solve costly and difficult problems relating to cryogenic space technology, because of the cryogenic propellant boil off problem. For these and other reasons, most U.S. orbit-to-orbit shuttle designs have avoided modular propellant tanks, and provisions for extensive refueling in space are practically nonexistent.

By 1970 the U.S. orbit-to-orbit shuttle was being designed with a fixed propellant capacity. Even though the Russian orbit-to-orbit shuttle rocket engines will be less efficient than the more sophisticated American rocket engines, U.S. mission studies showed that a smaller

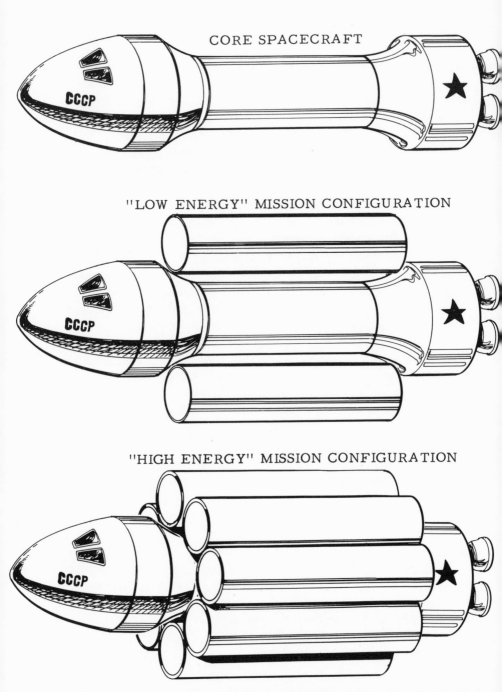

CORE SPACECRAFT

"LOW ENERGY" MISSION CONFIGURATION

"HIGH ENERGY" MISSION CONFIGURATION

PLATE 22 Artist's Concept of Russian Reusable Orbit-to-Orbit Shuttle—based on reliable information. (Modular propellant tanks offer Russians mission versatility.)

Russian spacecraft can still perform some space missions more efficiently than the American spacecraft. This is a point that many American space planners have consistently failed to grasp. In layman's language, this is another way of saying that even though a Volkswagen engine burns gasoline more efficiently than a Chevrolet engine, a Chevrolet powered by a Chevrolet engine can travel further on the same amount of gasoline than a Greyhound bus powered by Volkswagen engines. In other words, the emphasis in mission planning should not be on engine performance, but systems performance, so that all factors can be taken into account (i.e., the planners must consider engine performance, total weight of the vehicle, overall objectives, etc.). And this is what the design of the Russian orbit-to-orbit shuttle is based on, and why the Soviets do not get overly excited if they do not achieve the maximum possible engine performance from an engine, because the real issue is whether the entire space system—not just the engine—can get the job done.

By the end of 1970, U.S. defense planners learned that the key to the Soviet space effort hinged on (1) a space shuttle fleet that can deliver men, propellant, satellites, and spacecraft systems into orbit, (2) a network of orbital space stations that can receive men, equipment, and propellant, and (3) an orbit-to-orbit shuttle fleet that can perform complex orbital space missions, such as relay-type space missions and refueling in orbit, just as one would drive across the United States by periodically refueling at gasoline stations.

Because the Air Force's orbit-to-orbit shuttle would be carried into orbit inside the payload compartment of NASA's space shuttle, close coordination between the two organizations seemed like a logical requirement. U.S. intelligence analysts studying both the Russian and American efforts noted, however, that the American orbit-to-orbit shuttle design studies were unorganized, were not planned in the context of a total national space program, and were not being adequately coordinated with NASA. During this period I conducted mission studies on the U.S. orbit-to-orbit shuttle system because Pratt & Whitney Aircraft hoped that the Air Force would select their cryogenic rocket engine concept for use in the system. Because I was also studying the Russian space program at the time, I was not only familiar with the Russian approach, but I was in the position to confirm the charges made by other engineers that some Air Force planners over-specified the orbit-to-orbit shuttle rocket engine design requirements. U.S. rocket engine contractors were required to design the smallest and most sophisticated rocket engine possible that could deliver the maximum engine performance. Just as was the case with our launch vehicle design requirements, the emphasis was again on maximizing component performance instead of overall mission performance. As an example, the payload bay of the space shuttle's orbiter stage was specified by NASA to be 60 feet long, yet the Air Force's orbit-to-orbit shuttle engine design specifications were so stringent a one-inch deviation in rocket engine length could make or break the overall mission. These outrageous and unreasonable design specifications had engineers climbing the walls, and resulted in the proposal of costly exotic engines that used unorthodox collapsible nozzles and engines that required sophisticated mountings and special materials for the orbit-to-orbit shuttle system. Part of the reason for this was that the Air Force would not seriously consider extensive refueling operations in space, and for most missions the orbit-to-orbit shuttle had to be returned to earth intact. Thus, while the Russians planned to launch their orbit-to-orbit shuttles and leave them in orbit indefinitely, to be serviced and used periodically, the United States, because it was committed to use supercold cryogenic propellants that can easily boil off, and for other reasons, planned to return its orbit-to-orbit shuttles back to earth for launch again. Numerous U.S. engineers questioned the value of this approach, especially since it imposed completely unreasonable design specifications on the systems' rocket engines. Because of budgetary constraints, U.S. space and military planners were thinking on a near-term basis, and the frugal space missions planned by the Air Force and NASA could not possibly consider the problems that would occur if one planned a massive long-range assault on space, like the Soviets. And it made one wonder whether our space planners realized or even cared about the military significance of conducting operations in space. To further compound the confusion, some aerospace engineers privately criticized the Air Force because the managers of the orbit-to-orbit shuttle effort could not make up their minds and did not know what they really wanted. When things reached a low ebb in early 1971, and with further layoffs in the aerospace industry, the Air Force Rocket Propulsion Laboratory, the organization that was managing the orbit-to-orbit shuttle rocket engine program, encountered serious budgetary problems. Because of political pressure, their orbit-to-orbit shuttle rocket engine program became a make-work project to keep everyone busy; the idea of coordinating from the national level was the furthest thing from their minds because they had more important problems—the survival of the Air Force Rocket Propulsion Laboratory, which has served our country well in

148

the past, and was responsible for pioneering the high chamber pressure rocket technology that will be used in the space shuttle main engine being developed for NASA by Rocketdyne.

The gross lack of coordination between NASA and the Air Force reached its peak during the spring of 1971 when NASA, also because of budgetary problems, considered a smaller size payload bay for the orbiter stage of the space shuttle. This decision was made independently of Air Force considerations and it clearly showed that the two organizations were off and running in different directions, to the detriment of the American taxpayer and our national security. For all practical purposes the NASA recommendation to consider a smaller space shuttle payload bay eliminated the orbit-to-orbit shuttle system from further consideration, because the orbit-to-orbit shuttle could no longer fit in the space shuttle's payload bay. A space shuttle system without the orbit-to-orbit shuttle is about as good as a tractor trailer out of gas on the Los Angeles freeway. Shortly thereafter the Air Force complained and NASA agreed to adhere to the previously agreed upon payload bay dimensions. (The U.S. space shuttle's payload bay is shaped like a cylinder which is 60 feet long and 15 feet in diameter.)

The NASA-Air Force payload bay controversy also raised other questions. The payload bay of the Russian Rocketoplan is larger than that currently contemplated by NASA, and the reason for this is that the Russians characteristically think in terms of large bulky systems to minimize costs and to accommodate their anticipated military requirements in the decades to come. In layman's language, the problem would be equivalent to the following:

> A man buys a $100,000 unfurnished home. While his wife is out shopping for furniture, the man does not like the doors and windows in the home, so he replaces them with units that are half the original size. When the furniture is delivered, they find that none of the furnishings can fit through the door. The owner of the furniture store offers a solution; for additional cost, he can build tables, chairs, desks, beds, and other units with collapsible parts. Depending on how much the homeowners are willing to spend, the furniture store can redesign everything to fit through the front door. Obviously, this is one solution to the problem, but it is an expensive one.

Because we have engineers in the United States who are willing to design sophisticated payload systems that can occupy the least possible volume regardless of the cost, we also have people who continue to recommend that we develop a space shuttle system with a small payload bay.

After the payload bay disagreement was settled, the Air Force and NASA subsequently postponed the development of the orbit-to-orbit shuttle altogether because funds for the effort were not available. As of this writing, North American Rockwell is studying orbital operations for the orbit-to-orbit shuttle. Because NASA is contracting the study, the system is now called the Space Tug, but plans for actually developing such a system have been deferred. During a joint Army, Navy, Air Force and NASA propulsion conference in New Orleans in November 1972, I learned that several analytical studies were being conducted in the United States on space tug systems that use storable propellants. It is my hope that public disclosure of the Soviet effort via this book encourages those space officials in key positions to study the merits of storable propellants in a U.S. space tug system in more depth.

In summary, the Soviets are aggressively pursuing a low-cost orbit-to-orbit shuttle system that will complement their space shuttle operations. The Soviet orbit-to-orbit shuttle will be used for reconnaissance and surveillance missions, to deliver and retrieve secret satellites in remote orbits, to inspect and destroy hostile satellites, and provide logistics support for spacecraft and space stations in remote orbits. The Soviet orbit-to-orbit shuttle system will employ highly reliable reusable rocket engines that operate on low-cost, nonexotic propellants. During 1969, or before, the Soviets secretly simulated a space refueling operation by transferring nitric acid and kerosene in space. This experiment was in support of their program to perfect the space handling characteristics of simple propellant combinations for refueling orbit-to-orbit shuttle systems in space. Soviet orbit-to-orbit shuttles will be capable of undertaking both simple and aggressive long-duration space missions and will rely primarily on modular spacecraft construction (i.e., the number of propellant tanks that surround the core spacecraft can be varied to accommodate specific mission requirements) and extensive refueling in orbit to achieve this objective. On the other hand, the United States' orbit-to-orbit shuttle program, currently called the Space Tug Program, has been characterized by sophisticated and sometimes unreasonable design requirements, and after a tremendous upheaval within NASA and the Air Force because of design problems, gross lack of coordination, and budgetary problems, its development has been deferred. The Soviets enjoy a lead of about three to five years in the development of a reusable orbit-to-orbit shuttle, their program is clearly managed by the military, and they are very much aware of the necessity for developing such a system. Table VI summarizes the manned orbital spacecraft chronology between both nations, and Table VII summarizes the current status of this effort.

SAS-29

UNITED STATES			SOVIET UNION
	1961	Vostok - first manned orbital mission (one man) ▮ in April 1961.
Mercury - first manned orbital mission (one man) launched in February 1962.	1962	
	1963	
	1964	Voskhod - first manned orbital mission (three me launched in October 1964.
Gemini - first manned orbital mission (two men) launched in March 1965.	1965	
	1966	
Ground test of Apollo spacecraft in January 1967 kills three astronauts (the Apollo fire).	1967	Orbital mission of Soyuz spacecraft kills cosmona during reentry in April 1967.
Manned Apollo three-man spacecraft launched and successfully recovered in October 1968; Apollo spacecraft designed specifically to support a manned lunar landing mission (i.e., Apollo not directly comparable to Soyuz because spacecraft are designed for different missions). (Apollo 8)	1968	Manned Soyuz spacecraft launched and successfull recovered in October 1968; Soyuz spacecraft desig specifically to support near-earth orbital space m Soviets proclaim that Soyuz is a combination spac space station; special features include a three-ma reentry module and a spherical orbital compartme conducting experiments. Soviets initiate serious design studies on a reusab Orbit-to-Orbit Shuttle.
President Nixon's Space Task Group recommends Space Tug (reusable orbit-to-orbit shuttle) for deployment in middle-late 1970's.	1969	Soyuz 4 and 5 maneuver, rendezvous, and dock in January 1969 with stated objective of testing and improving design of the Soyuz spacecraft system; cosmonauts participate in orbital experiment. Soyuz 6, 7, and 8 maneuver and rendezvous in October 1969 with stated objective of testing and improving the design of the Soyuz spacecraft syste orbital space experiment includes three Soyuz spa ships flying in formation; seven cosmonauts partie in orbital experiment, including Shatalov and Yeli from the Soyuz 4 and 5 missions.
Detailed Space Tug design studies conducted.	1970	Soyuz 9 orbital mission lasts almost eighteen days Development of "modular" reusable Orbit-to-Orbi Shuttle initiated.
NASA defers development of Space Tug because of budgetary problems.	1971	Soyuz 10 launched in April 1971 for more tests and dock with Salyut space station; Cosmonauts Shatalc and Yeliseyev return to space in Soyuz for third ti Soyuz 11 launched in June 1971; three cosmonauts v Salyut space station and stay in orbit for almost 24 cosmonauts are killed during reentry due to decom pression of spacecraft from leak in spacecraft hat Soviets redesign hatch of Soyuz spacecraft.
(Apollo 17) End of Apollo Program	1972	Development of reusable orbit-to-orbit shuttle continues.

TABLE VI Manned Spacecraft Chronology

SAS-30

MANNED ORBITAL SPACECRAFT COMPARISON (PRESENT PLANS)

	US		TITLE HEADING	SOVIETS	
	Current Generation	Next Generation		Current Generation	Next Generation
	Apollo	Space Tug (Reusable Orbit-to-Orbit Shuttle)	NAME OF SPACECRAFT	Soyuz	Reusable Orbit-to-Orbit Shuttle
	45,000 pounds (earth orbital configuration)	up to 65,000 pounds	APPROXIMATE WEIGHT	14,500 pounds	20,000 - 100,000 pounds (based on modular spacecraft construction; number of propellant tanks varied depending on mission)
	1968	1985?	OPERATIONAL DATE	1968	late 1970's early 1980's
	After space rendezvous with Soviet Soyuz spacecraft in 1975, the U.S. has no further orbital spacecraft missions planned; Space Tug development program recommended by President Nixon's Space Task Group in 1969 deferred because of budgetary problems.		COMMENTS	Soyuz spacecraft will be used during 1970's to support manned space program; reusable orbit-to-orbit shuttle will be deployed when space shuttle becomes operational; orbit-to-orbit shuttle will complement space shuttle – space station complex and be used for reconnaissance and surveillance missions, to deliver or retrieve secret satellites in remote orbits, inspect and destroy hostile satellites, and provide logistics support for space stations in remote orbits.	

TABLE VII Manned Spacecraft Comparison (Present Plans)

SAS-31

Chapter 13

The Soviet Space Station Program

When the secret planners of the Soviet space program met to determine the course of their space effort, they decided that their troika of manned space systems (i.e., a reusable space shuttle fleet, an orbit-to-orbit shuttle fleet, and a network of orbital space stations) had to be developed methodically on a non-crash basis. This has become particularly true of the Russian space station program, which is outlined below.

The automatic rendezvous and docking of the Cosmos 186 and 188, and 212 and 213 unmanned spacecrafts in 1967 and 1968 represented the Soviets' first major step in developing the capability to erect orbital space stations without the participation of man. Then, in 1969, the successful rendezvous and docking of the manned Soyuz 4 and 5 spacecrafts represented the formation of the world's first mini-space station. In October of the same year the Soviets performed welding experiments in the Soyuz 6 spaceship, and they announced that these experiments represented another step towards the creation of large orbital space stations. Because Soviet cosmonauts would be required to conduct research and other functions in space stations for months on end, the Soviets needed physiological and psychological data to determine the effects of long duration space missions on man. In 1970, they obtained flight data from the crew of Soyuz 9, which stayed in orbit for almost eighteen days. Satisfied that they had enough information, their next step was to launch the world's first medium-size orbital space station in 1971. The station, called Salyut, weighed approximately 40,000 pounds, offered interior working space that was seventeen times greater than the Apollo spacecraft, and was fitted with sofas, chairs, and a dining table in addition to scientific and life support equipment. By comparison, the United States had two space station programs. The first, MOL (Manned Orbital Laboratory), was an Air Force project that was cancelled in 1969 because of budgetary problems. The MOL was originally scheduled to be launched in the late 1960s but delays pushed the launch date back to 1972, and when its overall objectives were compared to the technology available, it was concluded that MOL would not be as effective as other systems. For example, the original plan was to use the two-man Gemini spacecraft as the logistics system, but for the 1972 time period, this was a bit late; the three-man Russian Soyuz was operational in 1968 and it was infinitely more versatile than Gemini. Skylab is the NASA program that has converted the interior of the Saturn V moon rocket's last stage into a makeshift space station. Skylab, scheduled for operation in 1973, will be visited by three separate three-man crews that will stay in orbit for up to several months.

Salyut, which was placed into orbit in 1971 by the Proton launch vehicle, catapulted the Soviet Union into a commanding lead over the United States in the development of earth orbital space stations. (See Plate 23.)

The Soviets are currently developing their third generation orbital space station which will weigh in the 300,000 pound range, compared to 190,000 pounds for the American Skylab (Skylab weight figure also includes the command and service modules that are docked to the Skylab). The Soviet space station will be launched by the Super Booster, the launch vehicle that is larger than the Saturn V moon rocket. In general, during the middle 1970s we can expect the Soviets to conduct space operations using three types of space station systems: (1) docked Soyuz spacecraft (about 29,000 pounds of hardware; operations could involve up to six men); (2) Salyut (about 40,000 pounds of hardware; operations could involve up to eight or twelve men); and (3) a cylindrically shaped space station (about 300,000 pounds of hardware; operations could involve up to twenty men).

It is the Soviet objective to acquire orbital flight experience with these systems while the U.S. space program is dormant and to incorporate the knowledge learned from these programs into the

PLATE 23   Russian Drawing of the Soyuz Spacecraft and the Salyut Space Station (foreground). (Novosti Press Agency Photo)

designs of advanced space stations for the 1980s. In the late 1970s and early 1980s, the Soviets will routinely launch advanced space station modules weighing between 75,000 and 100,000 pounds into orbit with their space shuttle. According to Soviet present plans, these modules will be erected in space by manned and automatic means to form a massive space station complex, made up of medium-size, large, and super large stations. One space station design on the Soviet drawing boards will involve the link-up and welding of station modules to form the outer rim of a wheel. Other modules will form the spokes of the wheel, and when the station is rotated the cosmonauts who work in the outer rim of the station will have the benefit of centrifugal force as an artificial gravity to perform their work easier. (See Plate 24.) For specialized space research experiments or long-duration space missions that require both vacuum and zero-gravity working conditions, the Soviets plan to deploy individual space station modules that would be satellites of the main space base complex. These plans are aggressive, but they have been approved by the ruling members of the Politburo under the guidance of the U.S.S.R. Academy of Sciences.

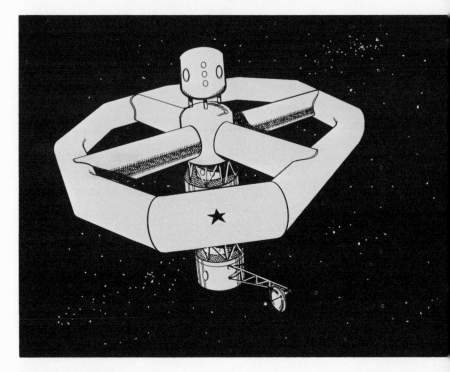

PLATE 24 Preliminary Plans: Russian Space Station for the 1980s—based on reliable information

In retrospect, Soviet interest in developing earth orbital space stations stems from their desire to exploit space because of scientific, economic, and military reasons. The Russian space program is wholeheartedly endorsed by the Soviet public and their cosmonauts are literally worshiped. One of the reasons for the pro-space euphoria that has engulfed the Soviet Union, besides its obvious propaganda value as a showcase for the world Communist movement, is the massive amount of literature on the space program that is made available to the Soviet public. Leading academicians, scientists, and engineers continuously write articles lauding the benefits of their space program, and because the press is controlled by the state, these articles frequently appear on the front page of Tass, Pravda, and Isvestiya. In the United States it is not uncommon to find a small paragraph about our space program on page 6 in section B of a third-rate newspaper. The public can thank the television news media, which did its share to blame the space program for the ills of our nation, for covering the last Apollo mission with reporters who thrived on talking about other national priorities. While it is not my intention to advise newspaper publishers and the television industry how to run their business, the public must recognize that the relative merits of our space effort is highly dependent on the public support it receives, and a disinterested press contributes to the weakening of our technological base, and ultimately, our national security. I have excerpted sections of a 1969 article by a Russian named Zaitsev, who in my opinion did an outstanding job of providing the Soviet public with the non-military reasons for building orbital space stations. I believe that the reader might find it interesting to see how the Russians justify the development of orbital space stations. The excerpts from the Zaitsev article are representative of the deluge of other Russian articles on their space program.

Condensed Excerpts From Zaitsev's Article*

Before we embark on voyages to other planets, there are a great many problems to be solved in closer proximity to Earth. One will be the establishment of orbital manned space research laboratories functioning over long periods of time.

Soviet Report, Center for Foreign Technology, Pasadena, California (No. 9, 1969).

The piloted orbital stations of the future are still very much on the drawing board. But one can already visualize more uses for them than it is possible to enumerate. Let us then confine ourselves to a few of the most likely applications awaiting them:

1. Space photos could be used to work out scientific schedules for

planting and harvesting crops, to pinpoint areas infested by pests in good time, to draw up crop forecasts, rationally to distribute farm machinery and transport facilities, to appraise the condition of grazing lands and fodder prospects on a countrywide scale.

2. A photographic record of the shifting of the snow and ice line could help to forecast floods.

3. Observation from outer space could be invaluable in the study and utilization of the wealth of the seas and oceans.

4. Surveys made from orbit could also help to trace ocean currents, draw up temperature charts of the seas and oceans, judge of the concentration of plankton by the color of the water to guide fishing fleets, for where there are plankton there are also fish.

5. Photographic observations from orbit could considerably improve the utilization of forest resources, not to speak of spotting forest fires or the damage done by pests and plant diseases.

6. Space laboratories could be useful for studying the Earth's crust, its composition and irregularities, and in revealing mineral deposits through measurements of gravitational and magnetic anomalies.

7. Orbital stations can quickly and precisely establish the geographic coordinates of remote points, ascertain the location of objects on aerial photographs, and define the contours of islands, reefs, shallows, and the like.

8. The processes taking place in the atmosphere have a most direct effect on life and human activity on Earth. It is a long-known fact that the radiation energy of the Sun is the prime mover of practically all physical and chemical processes on Earth and in its atmospheric envelope. The Sun constantly influences the magnetic field of the Earth as well as the atmosphere and the oceans. It pours down an endless stream of electrons and atomic nuclei, and this bombardment affects the ionosphere and the weather, causing such phenomena as the polar luminosity and magnetic storms. Study of all this could be greatly expanded by orbital stations.

9. Observations from space make for precise forecasting thanks to more exact tracking of cyclones, atmospheric fronts, and cloud formations. Eventually the genesis of cyclones will no doubt be revealed and it will become possible to influence the weather to one or another extent.

10. Orbital stations could conduct systematic investigations of the upper atmosphere to obtain a dynamic picture of the temperature and pressure in its various layers as well as of the chemical composition at different altitudes.

11. The magnetic field of the Earth will unquestionably remain a permanent object of study for orbital laboratories.

12. The solar ultraviolet, x-ray, and gamma radiation can be effectively investigated only outside the bounds of the atmosphere.

13. It is not excluded that the near future will see unique astronomical observations established on board stations orbiting the Earth. Such observatories would be guaranteed against atmospheric interference, the caprices of the weather and interference by the movement of masses of air.

14. Scientists have not yet been able to establish the source of the

mysterious cosmic x-radiation first discovered some ten years ago. Eventually telescopes carried out into outer space will solve this riddle.

15. Study of interstellar matter by spectroscopic observations of ultra-violet radiation will most likely afford an idea of its intergalactic density.

16. Radioastronomical observations from orbit would broaden our knowledge of the distribution of radiowaves of different lengths in interplanetary space. This would be of major practical value to the designers of communications systems for space devices traveling to vast distances from Earth.

17. Biological laboratories are another possibility. In particular, a study could be made of man's adaptability to those space flight conditions which cannot be artificially compensated, and experiments conducted to reveal the genetic effects of the cosmic environment.

18. Orbital stations offer attractive vistas to physicists as well. Outer space is packed with high energy particles with charges running into millions of billions of electron volts. The best the world's biggest accelerator can do is 70,000 million electron volts.

19. It is extremely difficult to achieve high vacuums, high energy radiation, extremely low temperatures, or powerful magnetic fields at the Earth's surface. The ideal place to do this is the orbital station, where nature provides all the conditions. Here, completely new technologies could be carried on in orbit, in many cases more efficiently than artificially created analogous conditions on the Earth's surface.

20. Lastly, orbital stations will be, in a way, a connecting link between Earth and other planets. They could help to solve many complex problems relating to the designing of ships and the training of crews for flight to other planets of our solar system. This applies in particular to technical verification of designs, elaboration of separate systems, and testing of new types of engines. Here spacemen could undergo acclimatization before flights. And the flights themselves could start not from the ground but from orbital platforms.

These are not the only possible uses. But they suffice to give an idea of how much orbital stations could contribute to both science and the economy.

By deeds and words the Soviet Union has demonstrated that the creation of orbital space stations is high on their priority list. The words of President of the Academy of Sciences Mstislav V. Keldysh and General Secretary Leonid Brezhnev summarize the Soviet position on this matter:

M. V. KELDYSH: "There are different trends and standards. But I think that the creation of orbital flights of orbital stations is the highest standard yet."

L. I. BREZHNEV: "Soviet science regards the setting up of orbital stations with crews that will be relieved, as man's highway into outer space."

Tables VIII and IX summarize the Soviet space station program. In contrast to the apathy expressed by many Americans over the merits of space exploration, the Soviets believe that the creation of a network of space stations is their highway into outer space. The creation of orbital space stations is a national goal that is receiving the highest possible support from both the Soviet political and scientific-technical establishments. Brezhnev's highway into outer space will be created by a series of orbiting space stations spatially arranged to accommodate both relay-type missions and extensive refueling operations by reusable orbital spacecraft systems. The Soviets believe that (a) the creation of orbital space stations is scientifically, economically, and militarily justifiable, (b) state that space stations can be used as platforms for lunar and interplanetary missions, research institutes, and bases, and (c) predict that significant scientific and technological breakthroughs will be realized through space research.

SAS-32

SPACE STATION CHRONOLOGY

UNITED STATES		SOVIET UNION
	1967	Successful automatic rendezvous and docking of unmanned Cosmos 186 and 188 spaceships in October 1967; Soviets laud achievement as another step towards the creation of large orbital space stations.
NASA conducts detailed manned orbital laboratory definition studies using existing Apollo hardware; program called "Apollo Applications"; deployment planned for early 1970's.	1968	Successful automatic rendezvous and docking of unmanned Cosmos 212 and 213 spaceships in April 1968; Soviets demonstrate capability to erect space station without participation of man in space.
Air Force Manned Orbital Laboratory (MOL) program cancelled because of budgetary and technical considerations. President Nixon's Space Task Group recommends space station program for deployment in middle-late 1970's; NASA initiates comprehensive space station program definition studies. NASA Apollo Applications Program design and development work continues.	1969	Successful rendezvous and docking of manned Soyuz 4 and 5 spaceships in January 1969 to form world's first mini orbital space station (about 29,000 pounds). Space welding experiments performed in Soyuz 6 launched in October 1969; Soviets laud achievement as another step towards the creation of large orbital space stations.
NASA continues detailed space station program definition studies. NASA Skylab (formerly called Apollo Applications) design, development, and ground testing work continues.	1970	Two-man crew of Soyuz 9 launched in June 1970 sets duration record of almost eighteen days in orbit; Soviets obtain physiological data applicable to orbital space stations.
NASA defers development of space stations because of budgetary problems.	1971	Launch of Salyut - world's first medium-size orbital space station (about 40,000 pounds) - in April 1971; three-man crew of Soyuz 11, launched in June 1971, visits Salyut for almost twenty-four days, setting another long-duration orbital space record.
NASA Skylab work continues.	1972	Soviets prepare for launch of second Salyut space station; significant work on 300,000 pound-class orbital space station is initiated.

TABLE VIII   Space Station Chronology

US	TITLE HEADING	SOVIETS
First Generation		**First Generation**
ab; uses redesigned last stage of rn V moon rocket.	NAME OR DESCRIPTION OF SPACE STATION	Docked Soyuz spaceships
000 pounds	APPROXIMATE WEIGHT	29,000 pounds
	OPERATIONAL DATE	1969
nched using Saturn-Apollo hardware.	SPECIFIC COMMENTS	Launched by operational Standard launch vehicle.
Second Generation		**Second Generation**
indrical modules assembled in orbit rm space stations; each module to be ched into orbit by space shuttle.		Salyut
o 65,000 pounds per module		40,000 pounds
's		1971
elopment plans deferred because of etary problems.		Launched by operational "Proton" launch vehicle.
		Third Generation
		Shaped like a cylinder
		300,000 pounds
		Middle 1970's
		To be launched by Soviet "Super Booster"
		Fourth Generation
		Cylindrical modules assembled in orbit by manned and automatic means to form space station complex.
		75,000 - 100,000 pounds per module
		Late 1970's early 1980's
		To be launched by first generation Soviet space shuttle.

GENERAL COMMENTS

Soviet space station program will continue to evolve methodically; Salyut and third generation space stations will be launched using existing launch vehicles to give the Soviets a viable manned space program during 1970's, while their space shuttle is being developed. When the space shuttle is operational, the Soviets will deploy fourth generation space station modules to form a network of earth orbiting space stations.

United States' space station development program for late 1970's and early 1980's has been deferred because of budgetary problems.

TABLE IX   Space Station Comparison (Present Plans)
SAS-34

To summarize, the Soviets are developing three advanced manned space systems that should become operational in the late 1970s, and most certainly in the early 1980s:

1. An earth-to-orbit *space shuttle* system.
2. *Orbit-to-orbit shuttle* system.
3. *Orbital space stations.*

Both American and Russian studies have shown that a space shuttle system must be complemented with a reusable orbit-to-orbit shuttle because numerous missions are beyond the capability of the space shuttle. Once Soviet planners decided to develop a space shuttle, their ministries, state committees, and design bureaus began to coordinate the effort accordingly. Changes in the payload specifications or dimensions of the payload bay of the Rocketoplan—the orbiter stage of the Russian space shuttle—would affect the design of their orbit-to-orbit shuttle, since the orbit-to-orbit shuttle must fit in the payload bay of the Rocketoplan; the size and usefulness of space stations is very dependent on the payload capacity of the space shuttle and orbit-to-orbit shuttle, and the cost of operating them. In contrast to the U.S. space effort, the Soviet space program is coordinated, planned, and supported from the grass roots and highest levels. It is conservative, methodical, well-planned, and emphasizes the use of existing and proven hardware. On the other hand, the United States is committed to develop only a reusable earth-to-orbit space shuttle system. Its effectiveness is very dependent on the development of a reusable orbit-to-orbit shuttle and orbiting space stations which the United States is not committed to develop, primarily because of budgetary reasons. American space planners continue to use the piecemeal approach to conquering space, and on a long-term basis this approach is costlier; it places a heavy burden on the taxpayer, and it is dangerous to our national security.

The current Soviet space effort, relative to ours, can be compared to the early 1950s, when we were designing launch vehicles that could deliver pounds of payload into orbit, while they were designing the standard launch vehicle which can deliver tons of payload into orbit. The fact that their space shuttle is being designed to deliver considerably more payload into orbit than the American space shuttle indicates that the Soviets are planning at least twenty years into the future again. The fact that they are pursuing the development of an extremely versatile, yet low-cost, orbit-to-orbit shuttle system, based on technology that they fully understand, while the United States has deferred the development of such a system, means that during the 1980s, the Soviets will be conducting routine, but ambitious near-earth military and civilian space missions. For the most part, Americans will be sitting on the sidelines, watching. This will become apparent in the late 1970s to those Americans who enjoy sitting in their backyards in the evenings looking at the stars. When groups of "stars" slowly move across the sky—perhaps a fleet of manned Soviet orbital spacecraft or a network of medium-size orbital space stations—concerned Americans who witness such happenings will ask: what could the Russians be doing up there? If this question is asked in the late 1970s, instead of today, then all could be lost; just as Japan could not respond in time to counter our technological breakthrough —the atomic bomb—we might not be able to respond in time to the development of a new revolutionary space-based weapon system, or one derived exclusively from extensive orbital space research.

Soviet space stations will play the same role in space that American aircraft carriers currently play on earth; Soviet orbit-to-orbit shuttles will play the same role in space that American fighter aircraft currently play on earth. The public should know that the Soviets are well on their way toward achieving superiority in space by 1980. (See Plate 25.)

SAS-35

Chapter 14

The Soviet
Strategic Threat

To place the Soviet space effort in its proper perspective, one must not only compare it against the U.S. effort, but it must be related to their overall defense program. In this section we examine Soviet political-military strategy, their strategic weapons systems, how they expect to use the systems developed, and the strategic threat to the United States.

The most authoritative unclassified Russian document that I have read on Soviet political-military doctrine was published in 1968 by the Military Publishing House in Moscow. This 464 page document, *Military Strategy*, was edited by the late Marshal of the Soviet Union V. D. Sokolovsky. It was approved for publication by the Central Committee of the Communist Party of the Soviet Union, and sold to Soviet citizens for two rubles (a little over two dollars). The paragraphs below contain excerpts from *Military Strategy*:

• "In describing the essence of war, Marxism-Leninism uses as its point of departure the position that war is not an aim in itself, but rather a tool of politics."

• "Politics prepares war and creates, for the benefit of strategy, favorable conditions in the economic and ideological respects."

• "The acceptance of war as a tool of politics determines the relation of military strategy and politics, based completely on the dependence of the former on the latter."

- "The main attention of military strategy is directed to studying the conditions under which a future war may arise, a detailed study of the peculiarities of the strategic deployment of the armed forces, the methods of delivering the first strike and conducting the first operations, as well as the method of strategic utilization of the different services of the armed forces."

- "The preparation of foreign policy for war includes such measures as the signing of treaties, the formation of coalitions, the safeguarding of the neutrality of neighboring countries, and others."

- "It is important for military strategy to assure the neutrality of a number of countries or of individual countries; this task is also assigned to diplomacy."

- "The wars between states with different social systems, the highest form of class struggle, are particularly decisive. In wars between states with the same social system, when there are no social contradictions between the antagonists, the political and strategic aims, the experience of imperialist wars show, are usually limited. In such wars, long before economic and military exhaustion of the belligerent states is reached, compromises of various types are possible."

- "The nature of military strategy is often influenced by such factors as the general historical, national, and political traditions of a country. For instance, Britain in its foreign policy always adhered to a clearly pronounced policy of watchful waiting, over-safeguarding, having someone else do their dirty work for them. This influenced their military strategy, which avoided decisive engagements, refused to take even reasonable risks, and always looked for devious, indirect roads to victory."

- "It is apparent that when the very outcome of the war depends largely on the number and the effectiveness of the strikes at the very beginning of the war, it is hardly reasonable to count on the potential capabilities of a country and to reserve a large part of the manpower for military operations during late periods of war."

- "Diplomatic and economic struggle does not stop in wartime, but these forms of political struggle are entirely dependent on the decisive form, that of armed conflict."

- "Under conditions of nuclear rocket war, the resolution of the main aims and problems of war will be accomplished by strategic rocket troops, by delivery of massed nuclear rocket strikes. Ground troops with the aid of aviation will perform important strategic functions in modern war; by rapid offensive movements they will completely annihilate the remaining enemy formations, occupy enemy territory, and prevent the enemy from invading one's own territory."

• "Consequently, overall victory in war is no longer the culmination, nor the sum of individual successes, but the result of a one-time application of the entire might of a state accumulated before the war."

• "Once the military operations on land and on sea have been started, they are no longer subject to the desires and plans of diplomacy, but rather to their own laws, which cannot be violated without endangering the entire undertaking."

In summary, *Military Strategy* and other official Soviet publications show that the Communist Party of the Soviet Union adheres to the doctrine that war is a tool of politics. On the other hand, most Americans believe that politics and diplomacy are tools for avoiding wars. Again, the Soviets note that in preparing for war, diplomacy should be used to negotiate treaties, coalitions, and the neutrality of neighboring countries—something that Hitler did before World War II. Yet most Americans regard diplomacy, negotiations and the signing of treaties as positive measures for creating conditions that will ensure "a generation without war, a lasting peace." And finally, many Americans and Western military theoreticians think in terms of avoiding a nuclear war at all costs, because if there were a nuclear exchange between the superpowers then everyone would lose; all politics would come to an end and universal mutual annihilation would begin. Marshal Sokolovsky disagreed with this view: "It is quite evident that such views are a consequence of a metaphysical and antiscientific approach to a social phenomenon such as war, and are a result of idealization of new weapons. It is well-known that the essence of war as a continuation of politics does not change with changing technology and armament." This is a significant statement because it is another way of saying that the Soviets find nuclear war acceptable if it can serve their political objectives (i.e., even with nuclear weapons, war is still a tool of politics). Military strategy of the Soviet Communist Party is predicated upon the fact that the success of a nuclear confrontation with the enemy requires the delivery of a massive number of nuclear rocket strikes and the one-time application of its entire strategic forces at the very beginning of the war. This means that the Soviets recognize the importance of maintaining a viable nuclear strategic force, and according to their political-military doctrine the more massive the force, the better. More significantly, the Soviets believe that once military operations begin they must not be interrupted by diplomatic measures, because this could endanger the success of the entire undertaking.

> Three points that should be made perfectly clear to all Americans are: (1) the Soviets believe that a war is a continuation of a policy by another means, (2) nuclear war is acceptable if it can serve the objectives of the Communist Party, and (3) treaties can be used to strengthen the position of the Communist movement.
> **SAS-36**

In spite of the SALT (Strategic Arms Limitations Talks) agreements co-signed by President Nixon and General Secretary Brezhnev in Moscow's Vladimir Hall in May 1972, U.S. officials have expressed great concern over the magnitude of the Soviet land-based intercontinental ballistic missile force. The SALT agreements were subsequently ratified by the U.S. Senate by an 88 to 2 vote, and further refinements and SALT treaties can be expected in the future. The story about the worrisome present situation begins in October 1962 during the Cuban missile crisis, when the United States greatly outnumbered the Soviets in land-based ICBMs (intercontinental ballistic missiles); in SLBMs (submarine launched ballistic missiles); and in long-range strategic bombers. When President Kennedy announced a blockade on the shipment of offensive weapons to Cuba and demanded that all offensive missiles be removed from the island, Premier Nikita Khrushchev acquiesed and complied with the President's demands. The question that most experts subsequently asked was: what would the Russians have done if they were strategically superior to the United States? The Russians learned from this humiliating experience never to negotiate or attempt to use force from a position of weakness.

After the Cuban missile crisis, the Soviets embarked on a massive undertaking to achieve military and nuclear superiority over the United States because the rules of world politics, and especially the Communist movement, favor the nation that is the strongest. Beginning in 1963–1964, under Khrushchev, the Soviets began building up their ICBM force. Department of Defense intelligence analysts following the Russian progress made periodic reports to the Johnson and Nixon administrations. In some instances their analyses were distorted by higher echelon intelligence officers, who incorporated their personal prejudices and beliefs into the finished intelligence assessments; these officials could not afford to drastically revise previously erroneous assessments because of job security reasons. Statements by the former secretaries of defense (McNamara, Clifford, and Laird) have been excerpted from their official annual reports to Congress, and are shown below. These statements generally depict their understanding of the Soviet ICBM arms build-up since the Cuban missile crisis.

1968—ROBERT S. McNAMARA

" ... Two significant changes have occurred during the last year in our projections of Soviet strategic forces. The first is a faster-than-expected rate of construction of hard ICBM silos. ... As of now, we have more than three times the number of intercontinental ballistic missiles (i.e., ICBMs and SLBMs) [than] the Soviets have. Even by the early 1970s, we still expect to have a significant lead over the Soviet Union in terms of numbers and a very substantial superiority in terms of overall combat effectiveness. In this connection, we should bear in mind that it is not the number of missiles which is important, but rather the character of the payloads they carry; the missile is simply the delivery vehicle."

1969—CLARK M. CLIFFORD

" ... We estimate that as of 1 September 1968 the Soviets had approximately *900* ICBM launchers operational, compared with 570 in mid-1967 and 250 in mid-1966—an increase of well over threefold in a period of a little more than two years. The rate of increase over the past year has been somewhat greater than estimated a year ago. However, we believe the rate of increase will be considerably smaller over the next two or three years."

1970—MELVIN R. LAIRD

" ... The projections for ICBM and SLBM strengths for mid-1970 and mid-1971 have been revised upward in each of the past five years as additional information on Soviet deployments has become available. ... The fact that our projections have not reflected all of the growth in Soviet offensive missile strength over the past several years is less important than the actual magnitude of this threat. ... The Soviets now have more operational ICBM launchers, over 1,100, than the United States, 1,054. More than 275 of these Soviet launchers are for the large SS-9 (ICBM). It is projected that there will be over 1,250 operational ICBMs on launchers by mid-1970. The change brought about by the Soviets in their strategic missile force is readily apparent when we recall that they had only about 250 ICBM launchers in 1966. At the current deployment rates, they will markedly improve the numerical advantage they already possess. In addition to quantitative increases, the Soviets are actively working on qualitative improvements, for example, their testing of multiple reentry vehicles with the SS-9."

1971—MELVIN R. LAIRD

" ... The primary strategic threat to the U.S.—the capability of the Soviet Union to deliver long range, nuclear weapons against targets in the United States—has been a matter of grave concern to us. ... The Soviets have built up their ICBM forces at a rapid rate during the past five years, and as of the end of 1970, had some *1,440* operational launchers. There are indications, however, that construction on new silo starts has slowed during the past year ... the explanation may be that the Soviets are preparing to deploy new ICBM systems."

1972—MELVIN R. LAIRD

" ... It is evident that the Soviets have built up their ICBM forces at a rapid rate during the past five years. As of the end of 1971 they had some *1,520* operational launchers ... during the course of the last year we detected almost 100 new silos that differ from currently deployed Soviet ICBMs. The implications of new silo construction are not yet completely clear, but the Soviets may be preparing to deploy two new or modified ICBM systems."

Plate 26 summarizes the Soviet ICBM build-up relative to that of the United States. The chart shows the number of operational launchers for each country since the Cuban missile crisis. In a period of just ten years, the Soviets reversed the strategic balance posture in numbers of ICBM launchers and currently have deployed about *1,620* ICBM launchers compared to 1,054 for the United States.

A strong case can be made that the Soviets gave up nothing when General Secretary Brezhnev signed the SALT agreement with President Nixon during the Moscow Summit of 1972. Just as the Paris peace talks on the Vietnam issue began during the Lyndon Johnson administration, and these talks lingered on for years, the Strategic Arms Limitations discussions were initiated on November 17, 1969, and after numerous meetings in Helsinki and Vienna, an agreement was reached in May 1972, when the Soviets decided that a 50 percent advantage over the United States in numbers of ICBM launchers would be sufficient.

The Soviet position on negotiations was best described by Dr. Richard Pipes, director of the Russian Research Center at Harvard University: " ... Soviet foreign policy adhers consistently to the

172

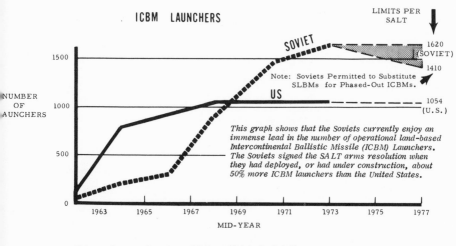

ICBM LAUNCHERS

LIMITS PER SALT

NUMBER OF LAUNCHERS

SOVIET

US

Note: Soviets Permitted to Substitute SLBMs for Phased-Out ICBMs.

1620 (SOVIET)
1410
1054 (U.S.)

This graph shows that the Soviets currently enjoy an immense lead in the number of operational land-based Intercontinental Ballistic Missile (ICBM) Launchers. The Soviets signed the SALT arms resolution when they had deployed, or had under construction, about 50% more ICBM launchers than the United States.

1500
1000
500
0

1963   1965   1967   1969   1971   1973   1975   1977

MID-YEAR

Primary Source - Secretary of Defense Melvin R. Laird's Annual Defense Department Report for Fiscal Year 1973.

PLATE 26   Number of Intercontinental Ballistic Missile (ICBM) Launchers Deployed
SAS-37

principle 'What is mine is mine, what is yours is negotiable.'" In Vietnam, for example, most Americans accept that North Vietnam will remain communistic, while the fate of South Vietnam is the negotiable commodity. Likewise, the Soviets built up an ICBM force that is second to none. After deploying about 50 percent more land-based ICBMs than the United States, Soviet leaders were willing to sign an agreement that will limit us from deploying additional ICBMs.

The American public was told that the United States could afford to sign the SALT agreement because (1) our missiles are equipped with multiple independently targeted reentry vehicles (i.e., MIRVs —warheads that can be independently directed to destroy different targets) while the Soviets do not have this capability, and (2) American warheads can hit their targets with greater accuracy than Russian warheads. On the near-term, these arguments are valid but they do not hold up on a long-term basis. The truth of the matter is that by the end of this decade the Soviets will possess multiple independently targeted reentry vehicles that can function just as accurately as those of the United States. However, the more numerous Soviet ICBMs are already larger than those of the United States, which means that they can carry a larger warhead, more individual warheads, more decoys to foil U.S. defenses, and more on-board protective materials to prevent them from being destroyed by U.S. weapons. And, because the Soviets have

173

deployed more ICBMs than the United States, when they develop a reliable MIRV capability—they are already working on it—the arguments of the past decade that the United States will still have a significant lead over the Soviet Union in terms of numbers of warheads will no longer be valid.

The United States has always emphasized the number of warheads in its arsenal rather than the number of delivery vehicles (i.e., number of ICBMs, bombers, nuclear submarines, etc). The Soviets on the other hand continue to stress the number of delivery vehicles. (See Plate 27.) A nation's first strike or retaliatory strike capability is dependent on both the number of delivery vehicles in its inventory and the number of warheads that these vehicles can carry. The point is: if the delivery vehicle can be destroyed before its warheads are deployed, then the weapon system can be rendered useless. And a nation's land-based delivery vehicles—ICBMs and bombers—can be destroyed by the launch of a preemptive first strike, or in the words of Marshal Sokolovsky: " . . . a one-time application of the entire might of a state accumulated before the war." It is Sokolovsky's logic—approved by the Communist Party—that prevails in the Soviet Union today, and this is one of the reasons why the Soviets have taken every conceivable measure to accumulate a massive number of ICBMs and delivery vehicles.

When the United States dropped the atomic bomb on Hiroshima, the weapon killed approximately 75,000 people and maimed countless

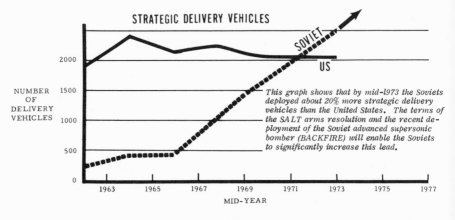

STRATEGIC DELIVERY VEHICLES

NUMBER OF DELIVERY VEHICLES

This graph shows that by mid-1973 the Soviets deployed about 20% more strategic delivery vehicles than the United States. The terms of the SALT arms resolution and the recent deployment of the Soviet advanced supersonic bomber (BACKFIRE) will enable the Soviets to significantly increase this lead.

MID-YEAR

Primary Source - Secretary of Defense Melvin R. Laird's Annual Defense Department Report for Fiscal Year 1973.

PLATE 27 Number of Strategic Delivery Vehicles Deployed
SAS-38

others in the holocaust. The destructive power of the first atomic bomb was said to be equivalent to the energy released from the chemical explosion of 20,000 tons of TNT, or 20 kilotons. With advances in nuclear weaponry since World War II, the explosive power or yield of nuclear weapons has increased significantly. A nuclear warhead from the Minuteman II ICBM in the U.S. arsenal, for example, carries the destructive power of about one megaton, which is equivalent to the destructive power of one million tons of TNT, or roughly 50 times greater than the Hiroshima atomic bomb. Each multiple independently targeted reentry vehicle warhead from the Minuteman III ICBM carries about ten times the destructive power of the Hiroshima bomb; the explosive power of these and other weapons varies, depending on the number of warheads deployed by the delivery vehicle. Most Americans have been told that the U.S. nuclear arsenal represents an overkill, many times over, but it is low by Russian standards. For example, the Soviet SS-9 ICBM can deliver a nuclear warhead to a U.S. target that yields 25 megatons, or as much as 1,250 times the destructive power of the Hiroshima bomb. The Soviets have also tested a multiple warhead version of the SS-9 ICBM, and it is known that this ICBM can deliver to U.S. targets three warheads that have an explosive yield of about five megatons each, or 250 times greater than the Hiroshima bomb. The Soviets have deployed over 300 SS-9 ICBMs; the destructive power of the Soviet SS-9 ICBMs alone exceeds the total destructive power of the entire U.S. nuclear arsenal. In blunt terminology, by the end of the 1970s Soviet ICBMs will be capable of annihilating the U.S. land-based ICBM force. *The point worth noting is that the Soviets can develop this capability without violating the terms of the SALT agreements.*

With the SALT dialogue continuing through the 1970s, if history is a guide for things to come, we should expect our leaders to make further concessions to the Soviets. This is the price that we will pay and have paid for just getting the Soviets to the conference table and for negotiating during an election year. A KGB intelligence officer, who is held in high esteem by the U.S. intelligence community because of his competence and shrewdness, once candidly told me: "We negotiate seriously with the United States only during election years, because we know that we get the best agreements then."

On the subject of SALT, the limitations agreed upon concern the number of ICBM *launchers*, not missiles. In the United States, land-based ICBMs have been designed on the basis of one missile per launch silo. The Soviets have deployed about 1,620 ICBM launch silos compared to our 1,054. Some Soviet ICBM launchers have been designed to be used over again. Just as a launch pad at Cape Kennedy is readied for reuse after a space launch, some of the Soviet ICBM silos have been designed for reuse; the turnaround time between launches for wartime conditions would be hours instead of days. Some U.S. experts believe that the Soviets are preparing for a first strike, and thus they plan to reload their silos for a second volley which can conceivably be used to either annihilate the remaining U.S. forces or blackmail the United States into surrendering.

Some Soviet SS-9 ICBMs have been equipped with orbital maneuverable space systems that can intercept and destroy American satellites in space. Other SS-9 ICBMs have been modified and can be used to inject the fractional orbital bombardment system (FOBS) into orbit. Whereas Western dialogue on this system has been directed to its inept first strike capability, the Russians believe that a modified version of the FOBS—a maneuverable orbital bombardment system that can be controlled from Moscow or orbital space stations—would also be valuable *after* hostilities break out, because once placed in orbit it could be used to coerce the President, or the commander of the remaining U.S. strategic forces, into surrendering.

In summary, the Soviets possess a first-rate land-based ICBM force that is greater in numbers and can deliver more megatonnage destructive power than existing U.S. ICBMs. By the end of this decade the Soviets will be capable of delivering multiple independently targeted warheads (MIRVs) to targets with about the same accuracy as U.S. ICBMs. Under these circumstances, Soviet ICBMs and submarine launched ballistic missiles (discussed in later paragraphs) will be capable of destroying the U.S. land-based ICBM force. Soviet ICBMs that have 1,250 times the destructive power of the Hiroshima bomb are *currently* capable of hitting U.S. targets which are within one mile of the predetermined impact point, and with improved accuracy they will be capable of destroying protected ICBM silos. The Soviet land-based ICBM force can support both a preemptive first strike against the United States, and provisions have been made for reusing ICBM silos and launching orbital space weapons for conducting operations after nuclear hostilities break out. Whereas some U.S. defense planners and politicians have consistently questioned the logic of developing a massive nuclear overkill capability, the Soviets do not share their view. All phases of Soviet ICBM force plans are in concert with their political-military doctrine that war is a tool of politics, and given the right conditions, a nuclear war can be won. Soviet ICBM force plans are based exclusively on their relative superiority in numbers over the United States; not on nuclear overkill sufficiency. The events of the past decade, in spite of the SALT agreements, show that the trends presented in General Bernard Schriever's (retired head of the United States Air Force Systems Command) July 1967 report to the House Armed Services Committee, predicting a significant deliverable megatonnage advantage for the Soviet Union over the United States during the early 1970s, are valid and can no longer be justifiably challenged. The Soviets currently enjoy nuclear superiority over the United States; their nuclear megatonnage arsenal is currently five times greater than ours, and the terms of the SALT agreements do not significantly impede the Soviet nuclear arms build-up over the next five years, despite talk to the contrary.

SAS-39

The Soviet sea-based ICBM force is formidable, and according to the conditions of the SALT arms resolution approved by Congress and signed by President Nixon on September 30, 1972, the Soviets will be allowed to deploy 84 missile firing submarines carrying 950 submarine launched ballistic missiles (SLBMs), compared to 44 U.S. missile firing submarines carrying 710 SLBMs for the United States. Thus, by the middle 1970s, the Soviets will have an appreciable advantage in SLBMs. Contrary to common belief, the SALT agreement does not impede the development of Soviet sea-based strategic forces, because to reach the upper limit of the agreement the Soviets must continue their arms build-up for almost five more years, and this is the point that Senator Henry Jackson adroitly made during the 1972 hearings on the SALT agreements. Plate 28 shows the number of operational U.S. and U.S.S.R. submarine launched ballistic missiles (SLBMs) since the days of the Cuban missile crisis. The upper limits of the SALT arms resolution are also indicated.

Since the Cuban missile crisis, the Soviets have built up a naval force that is second to none. Americans watching the evening news have become accustomed to the Soviet naval presence in the Mediterranean,

SLBM LAUNCHERS

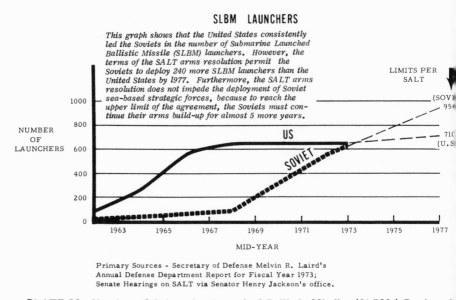

This graph shows that the United States consistently led the Soviets in the number of Submarine Launched Ballistic Missile (SLBM) launchers. However, the terms of the SALT arms resolution permit the Soviets to deploy 240 more SLBM launchers than the United States by 1977. Furthermore, the SALT arms resolution does not impede the deployment of Soviet sea-based strategic forces, because to reach the upper limit of the agreement, the Soviets must continue their arms build-up for almost 5 more years.

LIMITS PER SALT

(SOV)
95(

US

710
(U.S

SOVIET

NUMBER OF LAUNCHERS

1000
800
600
400
200
0

1963   1965   1967   1969   1971   1973   1975   1977

MID-YEAR

Primary Sources - Secretary of Defense Melvin R. Laird's Annual Defense Department Report for Fiscal Year 1973; Senate Hearings on SALT via Senator Henry Jackson's office.

PLATE 28   Number of Submarine Launched Ballistic Missiles (SLBMs) Deployed SAS-40

the Atlantic Ocean, the Caribbean off the coast of Florida, and the Indian and Pacific Oceans. In a historic speech delivered by the late Congressman L. Mendel Rivers to the members of the House of Representatives in September 1970, and subsequently documented in a significant 1972 publication entitled, *America Faces Defeat*, by General Lewis W. Walt, USMC (Ret.), it was shown that the U.S. Navy is no longer the greatest naval power in the world. Two significant points that General Walt made were:

(1) In 1972, the Soviets had over 400 submarines, compared to 135 for the United States; 98 Soviet submarines were nuclear powered compared to 93 for the United States. (Note: totals include ballistic missile, cruise missile, patrol, and attack nuclear powered submarines.)

(2) Two-thirds of the active U.S. fleet is over twenty-years-old, while only one-tenth of the active Russian fleet is over twenty-years-old.

It can be stated that the Soviets currently possess a first-rate navy that is second to none, and a sea-based strategic force that is a viable threat to U.S. land-based ICBMs and other strategic weapons systems. The 1972 SALT agreements have ensured the Soviets superiority in the number of sea-based weapons systems, and this is consistent with Soviet negotiating philosophy.

SAS-41

In a 1970 intelligence report to the CIA, I reported: "The NK-144 two-spool turbofan engine will be modified slightly and incorporated into a Soviet long-range bomber weapon system." This was another way of saying that the Kuznetsov Engine Design Bureau was modifying the engine they developed for the Tupolev 144 supersonic transport aircraft—the faster-than-sound civilian passenger aircraft —for use in an intercontinental bomber weapon system. In 1971 the Department of Defense publicly acknowledged that the Soviets were indeed developing such a system, which has subsequently been given the NATO code name BACKFIRE. BACKFIRE can travel about 1,500 miles per hour and carry nuclear bombs and air-to-surface nuclear missiles. The Soviets initiated production of BACKFIRE in 1972. By comparison, the U.S. long-range supersonic bomber weapon system—the B-1, being built by North American Rockwell Corporation—will not become operational until 1978–1980. Furthermore, when the B-1 long-range bomber becomes operational there is reliable evidence to indicate that it might not be as effective as BACKFIRE, not because of aircraft design considerations but because of the Soviet air defense network, made up of advanced missiles and interceptor aircraft. Some U.S. experts believe that the B-1 bomber will not be capable of penetrating the expected Soviet air defenses during the 1980s, just as some B-52 bombers could not penetrate the air space over North Vietnam. (See Plates 29 and 30.)

In summary, for the late 1970s and 1980s, the Soviets will have a first-rate nuclear threat made up of land- and sea-based ICBMs and long-range supersonic bombers. The terms of the SALT agreements permit the Soviets to deploy more and larger ICBMs than the United States, and they already enjoy about a five year lead over the United States in the development of a long-range supersonic bomber weapon system.

SAS-42

It is now clear that during the past ten years the Soviets have been very busy developing space and strategic weapons systems, and it has never been their intention to achieve nuclear parity with the United States, as was claimed by numerous U.S. government officials during the 1960s. Never before has a nation embarked on such a strategic arms build-up in such a short period of time. The alarming pace in the

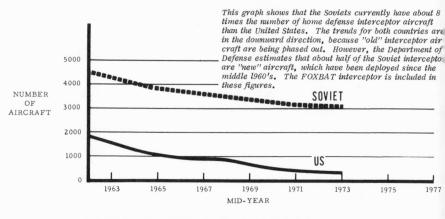

INTERCEPTOR AIRCRAFT

This graph shows that the Soviets currently have about 8 times the number of home defense interceptor aircraft than the United States. The trends for both countries are in the downward direction, because "old" interceptor aircraft are being phased out. However, the Department of Defense estimates that about half of the Soviet interceptors are "new" aircraft, which have been deployed since the middle 1960's. The FOXBAT interceptor is included in these figures.

Primary Source - Secretary of Defense Melvin R. Laird's
Annual Defense Department Report for Fiscal Year 1973.

PLATE 29   Number of Interceptor Aircraft Deployed
SAS-43

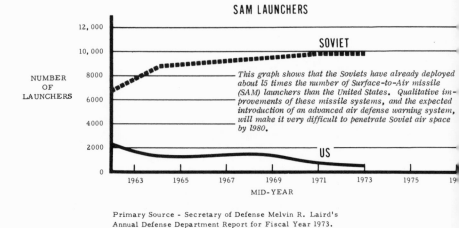

SAM LAUNCHERS

This graph shows that the Soviets have already deployed about 15 times the number of Surface-to-Air missile (SAM) launchers than the United States. Qualitative improvements of these missile systems, and the expected introduction of an advanced air defense warning system, will make it very difficult to penetrate Soviet air space by 1980.

Primary Source - Secretary of Defense Melvin R. Laird's
Annual Defense Department Report for Fiscal Year 1973.

PLATE 30   Number of Surface-to-Air Missile (SAM) Launchers Deployed
SAS-44

deployment of Soviet weapons systems is reminiscent of the Nazi arms build-up in the 1930s, and it makes one wonder if we are on the verge of World War III. If one considers the Soviet military-political philosophy, it is readily understood that war is a tool of politics, and the Soviets have developed weapons systems whereby they can impose their will on weaker nations, not through diplomatic negotiations but through the application of force. The continuity and single purpose of the Communist movement—to impose its will on the free peoples of the world—has remained intact over the years, as witnessed by the Communist takeover of the Eastern bloc satellite countries by Stalin after World War II, the invasion of Hungary in 1956 under Khrushchev, and the invasion of Czechoslovakia in 1968 under Brezhnev and Kosygin. After the Soviets invaded Czechoslovakia, many U.S. intelligence analysts credited the Soviet Politburo with a masterful tactical move, because it was the consensus of these analysts that the world would soon forget. They were correct: who would have believed, for example, that the President of the United States would participate in a Moscow Summit Conference less than four years after the invasion, or that the United States would bail the Soviets out of a massive food shortage crisis by selling them over one billion dollars worth of wheat at bargain basement prices? Insiders, who have studied the Soviet arms build-up and have followed the recent agreements and negotiations, have found the dialogue about a detente between East and West very disturbing, when it is known that a detente to a Westerner means a true relaxation in tensions with the prospects for peace greatly improved, while detente to Soviet leaders represents another tactical move on the world chessboard, designed to ultimately improve the Soviet strategic position. When one understands the principles of Soviet negotiating policy, one can then explain why in 1962 the Soviets abrogated the agreements they reached with the United States on the nuclear test moratorium; the Soviets needed nuclear test data applicable to developing an anti-ballistic missile defense system so they violated the terms of the moratorium to conduct a series of well-planned nuclear atmosphere tests. When their test program was completed, they signed the Nuclear Test Ban Treaty of 1963, which essentially prevented the United States from conducting similar atmospheric nuclear tests to catch up with the Soviets. As a result of the Soviet tests, and the subsequent test ban treaty, the Soviets have information on the effects of large nuclear explosions in the atmosphere, while U.S. data on this vital ABM-related information is limited.

When the Soviets signed the Non-Proliferation Treaty in June 1968—the treaty was regarded by Western "experts" as another genuine step to lessen tensions between the East and West—the Soviets invaded Czechoslovakia only months later. *The point that should be reiterated is that the Soviets use negotiations and treaties to improve their strategic position; each agreement with the United States can be thought of as the move on a chessboard, and above all, the agreement is independent of other considerations and problems.* Thus, on one hand, the Soviets signed the Non-Proliferation Treaty, creating an illusion of a detente with the West, and on the other, they invaded Czechoslovakia, arrested Alexander Dubcek, the party secretary of the Czech government, and declared the "Brezhnev Doctrine," which gives the Soviets the right to intervene in the affairs of any "socialist country" if the Soviets believe that the country is heading in a "non-socialist" direction.

Again, with President Nixon and Secretary of State Henry Kissinger proclaiming to the nation that the "new" detente with the Soviets was another step towards "a generation of peace," the Soviets moved clandestinely in October 1973 and assisted the Arabs in the planning and execution of preemptive strikes against Israeli fortifications in the Sinai and Golan Heights. The Yom Kippur War in the Middle East, fueled by secret Soviet promises to Egyptian leaders and arms shipments which included the Soviet Scud, a small surface-to-surface missile equipped with nuclear warheads and with a range of about 180 miles, showed the brazenness of Soviet foreign policy during an atmosphere of "detente." With the President of the United States beset with Watergate problems, Soviet military strategists moved swiftly. They hoped that the Yom Kippur War would incite the Arabs to cut off oil shipments to the United States and our NATO allies. Additionally, it was their objective to reopen the Suez Canal so the Soviet naval fleet could have immediate access to the Persian Gulf and Indian Ocean.

During the Cuban Missile Crisis, General Secretary Nikita Khrushchev was forced to remove offensive missiles from Cuba. A decade later, the superpowers confronted each other over the Middle East, and in spite of President Nixon's worldwide military alert, General Secretary Leonid Brezhnev has armed the Arabs with nuclear weapons, with advanced surface-to-air missiles, and with fighter aircraft such as the MiG-25 FOXBAT high altitude fighter—a move which has strengthened the Soviet military-political position in the Middle East.

The Soviets have not attempted to impose their form of government on the United States because of our deterrent capability. But as of this writing, the Soviets are strategically superior to the United States. And if the current trends continue, by the early 1980s the Soviets will not only be the most powerful nation in the world, possessing the most numerous and advanced nuclear and strategic weapons systems, but they could initiate a nuclear war and win it.

Chapter 15

World War III?

In the first chapter the following question was raised: how is it possible for the United States to spend approximately 80 billion dollars a year on defense and find its national security gravely threatened by a nation whose gross national product is about one-half as much? The answer is that money or the productivity of a nation does not buy defense. For years Americans have gone to bed each night believing that they are safe from a Communist takeover because of our superior strategic position. The preceding chapters have indicated that strategic superiority now belongs to the Soviets. Our security is dependent on how we expend our resources and how our strategic forces compare to the Soviets. Again, the chessboard analogy is appropriate. For those readers who are not familiar with chess, one can win by checkmating an opponent. A checkmate condition occurs when a player's king on the chessboard is threatened by his opponent's chesspiece(s), and even though the king is permitted to move to another square on the board to elude the threat, all possible escape routes and squares accessible to the king are also threatened. The game is therefore over and the king is checkmated. Chess differs from checkers because the king can be checkmated even though other chesspieces of the king's court (i.e., pawns, rooks, knights, bishops, and the queen) have not been eliminated. In other words, a player can lose in chess even if he has not lost

a single chesspiece to his opponent up to the time of checkn
player who loses such a game can be charged with not making e
use of the offensive and defensive forces at his disposal. On th
hand, a player can lose at chess because his opponent annihilate
or all, of the chesspieces at his disposal, thus leaving th
defenseless and vulnerable to a checkmate.

Just as in chess, the security of our nation depends on how effi
we utilize the offensive and defensive forces at our disposal vis-à-
Soviet Union; the magnitude of our space and defense expen   ..cs
contribute towards our defense, but they can never ensure it. The
Soviets have always recognized that we have a larger industrial and
economic base than themselves. We attribute this to our capitalistic
system, but KGB agent Nikolai Beloussov explained it to me as
follows: "We are a new nation! Only a little over fifty years. You have
been in existence for hundreds of years!" (I.e., the Communists start
their calendar with the Bolshevik Revolution in 1917, and they argue
that Communist nations will be more productive than capitalistic
nations, given the same opportunity or time to prove themselves.)
Because of our current superior industrial capacity, the Soviets have
elected to concentrate their military resources in areas that they can
successfully compete against us. Table X is self-explanatory and
summarizes the strategic force posture that we can anticipate during
the 1980 time period.

As Table X shows, by the end of this decade, unless the United
States changes its way of doing things, the Soviets will deploy more
ICBMs, SLBMs, advanced supersonic bombers, and space-based
weapons systems of comparable quality than the United States. Since
the Soviets are currently spending about three billion dollars a year
more than the United States on military research and development
programs, there exists the possibility that the more numerous Soviet
weapons which will be deployed in the future will be of better quality
than those of the United States. The weapons that currently contribute
to our national defense were developed from the technology available
to us in the early 1960s; conversely the weapons that we deploy in the
early 1980s will be based on the state of our technology today, and in
recent years the Soviets have spent more money and have invested a
greater percentage of their resources on military research and development programs than the United States.

The reader should reexamine Table X, because it is a graphic way of
saying that by 1980 our nation will become a second-rate power if the
current trends continue; the United States will have conceded supe-

STRATEGIC BALANCE

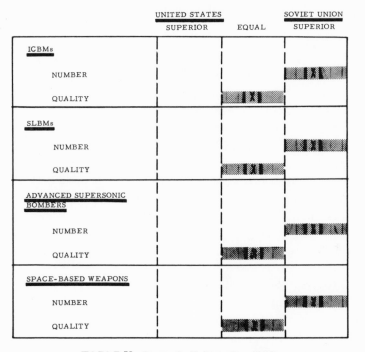

	UNITED STATES SUPERIOR	EQUAL	SOVIET UNION SUPERIOR
ICBMs			
NUMBER			▓▐▌█▐▌▐
QUALITY		▓▐▌█▐▓	
SLBMs			
NUMBER			█▐▌█▐▌▐
QUALITY		▓▐▌█▐▓	
ADVANCED SUPERSONIC BOMBERS			
NUMBER			█▐▌█▐▌█▐
QUALITY		▓▐█▐▓	
SPACE-BASED WEAPONS			
NUMBER			█▐▌█▐▓
QUALITY		▓▓█▐▌▐	

TABLE X Strategic Balance by 1980
SAS-45

iority to the Soviets in land-based intercontinental ballistic missiles
ICBMs), submarine launched ballistic missiles (SLBMs), and
ntercontinental bombers. Lest we forget, these are the three areas that
are the backbone of our offensive strategic forces. But, the Soviets have
added a new element to the equation—space. This is the weak link in
our planning structure. The Soviets know it, and they have been doing
something about it. Soviet planners recognize that they can checkmate
the United States into a secret surrender or win a decisive military
engagement if they complement their ICBMs, SLBMs, and advanced
supersonic bomber weapons with advanced space-based offensive and
defensive weapons. They also recognize that earth-based weapons
systems have their natural technological limitations, because in the
near future it will not be cost-effective to pour more money into the
refinement of existing strategic weapons. Though the military implications of man's thrust into space will not be fully understood for years
to come, just as the implications of air power were not fully understood

in the early 1900s even by our most astute military strategists, the Soviet methodical development of advanced space launch vehicles and space systems can no longer be ignored; the Communist Party does nothing without a purpose.

Based on the space systems currently under development in the Soviet Union, one can reliably forecast what the Soviets plan for the future. First, the Soviets will launch several more Salyut space stations and Soyuz spaceships before the long-awaited U.S.-U.S.S.R. Apollo Soyuz joint rendezvous and docking space mission in 1975. These missions will provide the Soviets with additional orbital flight experience applicable to long-duration space flights. After the fixes on the Soviet Super Booster are made, the Soviets will launch a 300,000 pound class earth orbital space station that can house up to twenty men. The standard launch vehicle, in conjunction with the three-man Soyuz spacecraft, will be used to ferry cosmonauts to and from the space station during the middle and late 1970s. The Super Booster will also be used to launch a manned expedition to the moon. In contrast to the Apollo program, Soviet plans call for the launch of two manned spaceships to the moon during the same launch opportunity; the second spaceship will serve as a lifeboat, should anything go wrong with the primary spaceship during the mission. When Soviet cosmonauts walk on the moon in the near future, Americans should recall the pronouncements of television network commentators during the December 1972 Apollo 17 mission, who ruled out manned lunar activity for the remainder of this century. And when the Soviets build the world's first lunar base only years later, the public should hold certain members of the U.S. news media accountable for their shortsightedness.

During the middle 1970s, U.S. intelligence sources and newspapers will report that the Soviets are conducting flight-tests with a reusable rocket-airplane, the Rocketoplan. The Rocketoplan—the orbiter stage of the Soviet space shuttle—will first be tested using an existing throwaway rocket booster, and for the most part the flight-test program will be supervised by the Soviet Academy of Sciences, Strategic Rocket Troops, and Air Force on behalf of the Ministry of Defense. The Soviets will also (1) initiate orbital flight-tests of a reusable orbit-to-orbit shuttle system and (2) experiment with mini-orbital space station modules that can fit in the cargo bay of the Rocketoplan. After the Soviets successfully test these systems (during the late 1970s), they will routinely launch orbit-to-orbit shuttle systems and space station modules into orbit using the Rocketoplan and an existing expendable ooster. The orbit-to-orbit shuttle system will use easy-to-handle orable propellants, and the spacecraft system will be capable of erforming both manned and unmanned space missions. At first, most anned missions will be conducted near space stations, and the Soviets ill rely on automatic systems to (1) deliver and retrieve satellites in mote orbits, (2) refuel spaceships in orbit, and (3) rendezvous and ock spaceships and space station modules.

By the early 1980s the advanced reusable airplane-type booster stage f the space shuttle should be completed, and according to current oviet plans it will employ both rocket engines and advanced jet ngines (scramjets). When this booster becomes operational and is ated to the Rocketoplan, the Soviets will have the only airplane-type ace shuttle in the world (i.e., both stages will be manned reusable ystems) and they will literally be capable of operating their space uttle like an airplane company operates its jet aircraft.

In the early 1980s, the United States should also be operating a space uttle, but because of budgetary, political, and technical decisions ade ten years earlier, the United States will be recovering the first age of its space shuttle in the ocean after each launch. For a onaggressive space program, such as that currently planned by U.S. ace officials and Congress, an ocean recovery space shuttle is cceptable. This, however, would not be acceptable to the Soviets ecause of military and economic reasons.

With NASA having fallen into the trap of trying to justify the space uttle program to the taxpayer in terms of benefits to mankind, the xpayer can expect the United States to plod along on the piecemeal ace program it initiated in the early 1970s, when it was common ractice to publicize the benefits of space by showing the man--the-street the value of photographing Farmer Brown's disease-dden cotton field in Georgia. This was the period when it was shionable not to alarm the public with dialogue on the military uses f space.

In the early 1980s, however, orbital space will be saturated with ractically every conceivable Soviet satellite, from passive to aggressive ace systems, from maneuverable spaceships which can seek and estroy orbiting U.S. satellites, to orbital weapons systems which can estroy terrestrial targets as well. The Soviets will have the capability to onduct both civilian and military operations in space on a routine asis, just as military aircraft are currently used in surveillance, connaissance, tactical, or strategic missions. During this period, the nited States will be limited by the aerospace hardware that it has

deployed. With some aerospace systems taking upwards of ten years t
research, develop, test, evaluate, and produce, the situation cannot b
corrected overnight.

It can be anticipated that as the Soviets accumulate operation:
experience with their space shuttle, routine Sunday afternoon launche
will become a common occurrence. As the Soviets flood the skies wit
orbiting space stations and spaceships, and conduct extensive scientifi
technological, and military orbital research, it can be expected tha
they will arm this network with advanced laser death ray weapons t
form the heart of an advanced Soviet antiballistic missile (ABM
defense network. Should the Soviets complement their massive earth
based nuclear arsenal with space-based weapons, and they are buildin
the hardware to do this, the stage would be set for a nuclear showdow
between the superpowers, and the odds would greatly favor the Soviet:
For example, the Soviets could launch a preemptive strike and destro
the majority of our land-based ICBM forces and strategic bomber:
The United States would deploy those long-range bombers and launc
those ICBMs and SLBMs that were not destroyed by the Soviet firs
strike. The combination of a Soviet space-based laser defense network
and a first-rate air defense network of surface-to-air missiles an·
interceptor aircraft could destroy a high percentage of the remainin
U.S. strategic forces; U.S. ICBMs and supersonic bomber aircral
would be shot down by laser weapons from orbiting space stations an·
spaceships above, and surface-to-air missiles and interceptor aircral
below.

It was no coincidence that the wording of the Outer Space Treaty i·
January 1967 did not preclude the use of laser weapons in orbita
space. By 1967, the Soviet government was already committed to botl
an aggressive near-earth space program and an advanced lase
weapons systems development program. They achieved a majo·
technological breakthrough in laser weapons technology in the earl:
1960s, and by the middle 1960s, they built a super secret complex in th·
steppes of Siberia to research and develop lasers. Since then, U.S. lase
researchers have been trying to catch up to their Soviet counterparts, a
demonstrated by the secret work and tests conducted by U.S
researchers at Kirtland Air Force Base outside of Albuquerque, Ne\
Mexico. While much has been said of U.S. successes in laser weaponry
let it be known that it was common for U.S. laser researchers to rely o·
Soviet textbooks and publications during the late 1960s as bibles i·
laser technology—to the point that some Russian textbooks an·
publications which could be purchased in a Washington book stor·

were classified in official government circles. While the vague terms of the SALT agreement reached in Moscow in 1972 preclude the use of ABM systems in space, the laser weapons programs in the United States and the Soviet Union still have A-1 priority, and scientists from both countries are investigating the uses of lasers to earth and space-based ABM systems.

In preparation for the eventual deployment of space-based laser weapons, the Soviet espionage network, the KGB and GRU, has been quite busy in providing the Soviet defense establishment with strategic and tactical information about U.S. defense installations and materials used in U.S. aerospace systems and nuclear warheads. It is known, for example, that lasers can be used to neutralize or detonate nuclear warheads, and the more powerful the laser beam, the more effective it is. During the early 1970s, U.S. analysts determined that the Soviets were actively engaged in researching lightweight structural materials such as inflatable structural elements and transparent plastics for use in advanced orbiting space stations. In a joint effort with the East Germans, Soviet researchers studied the adhesion characteristics of these materials to steel, aluminum, titanium, and copper—materials for various space-based systems. Top Soviet researchers at the Institutute of the Problems of Mechanics in Moscow and other Soviet research institutes studied the interaction of laser beams with these materials. Most of the research was directed at understanding the destructive mechanism of why and how laser beams destroy materials. The research was extended to determine the interaction of laser beams on target materials which were coated with silver and mirrored surfaces. Because a mirrored surface tends to reflect a laser beam, just as light is reflected by a mirror in one's home, the structural material beneath the mirrored surface can be protected. However, when the intensity of the laser beam is increased, or for other technical reasons, the beam can penetrate the protective mirrored surface and destroy the material beneath. The purpose of the Soviet tests was to investigate methods of protecting space-based structural materials from laser weapons. While there is a great amount of work that must be done in this area, because there are countless types of laser weapons and space applications, U.S. analysts have expressed admiration for the quality of the Soviet work in this area.

Besides near-earth space, planners from the Ministry of Defense have considered the use of libration regions for military purposes. Libration regions, located about nine-tenths of the way to the moon, are regions in space where the earth's gravitational force attraction is counterbalanced by the moon's gravitational force attraction. Thus, spacecraft parked in a libration region could remain there indefinitely as if suspended in space. In a libration region, astronauts would still experience zero gravity—they would float about in their spacecraft —but their spacecraft would be essentially motionless, as seen from the earth or moon. Libration regions can be used as parking spaces for military space systems; weapons could be stored for years and deployed immediately upon command. Space weapons stored in libration regions would be most valuable as second strike or retaliatory response weapons. For reasons mentioned in earlier sections, propulsion systems that use storable propellants would be ideally suited for these applications, since the space system would not have to be serviced for years. Some U.S. planners believe that the use of libration regions for weapons storage would be wasteful and inefficient. But again, these are generally the same experts who have argued against a fractional orbital bombardment system (FOBS), a maneuverable satellite capability, an orbit-to-orbit shuttle, and other programs which the Soviets believe are very worthwhile. If one applies Soviet standards and logic to the problem, the use of libration regions in space to store weapons and space systems would be a very sound investment because the Soviets believe that their strategic capability should not depend entirely on earth-based weapons systems in time of war. As Marshal Sokolovsky stated in *Military Strategy*: " ... Overall victory in war is no longer the culmination, nor the sum of individual successes, but the result of a one-time application of the entire might of a state accumulated before the war." In Soviet parlance, this includes a massive number of (1) ICBMs, (2) SLBMs, (3) advanced supersonic bombers, (4) maneuverable FOBS, (5) maneuverable satellite killers, (6) weapons stored in libration regions, (7) earth- and space-based laser weapons, (8) space shuttles, (9) orbit-to-orbit shuttles, (10) space stations, and (11) other aerospace systems too numerous to mention.

The current anti-space euphoria that has engulfed the United States is reminiscent of the period when General Billy Mitchell attempted to impress the U.S. military establishment with the importance of airpower in conducting military operations. Likewise, numerous U.S. officials in positions of authority have reacted negatively to the new frontier—space—yet the adverse effects of their decisions will not surface for years to come. Those who have studied the Soviet space effort have seen the Russians pursue this frontier with vigor, while the United States has rested on the laurels of our past achievements. Must we literally wait for the Soviets to flood the skies with spaceships and pace stations before we react? Just as automobiles, trucks, ships, rains, and airplanes have assisted underdeveloped nations into making he best possible use of the resources at their disposal, the aerospace ystems currently being developed within the Soviet Union will permit heir scientists, engineers, military planners, and strategists to exploit pace in the years ahead. Members of the U.S. intelligence community re gravely concerned over what the new Soviet aerospace systems nean in terms of the strategic balance, and because of the unknown ... he possibility of a technological Pearl Harbor.

Unfortunately, we do not have a guarantee that our Republic will urvive as many years as the Roman Empire; advances in technology ave made it possible for a nation to be destroyed in seconds, rather han years. It is for this reason that we should be aware of the Soviet pace and strategic capabilities, because there is no room for error.

Chapter 16

What Can <u>You</u> Do?

The machinery does not exist in the United States where intelligence analysts can be candid with their superiors without placing their jobs in jeopardy. The machinery does not exist where senior intelligence analysts can prepare intelligence assessments for the United States Intelligence Board without first modifying their reports to be fairly consistent with previous assessments, regardless of their validity. The machinery does not exist where the director of the CIA can advise the President without considering the political repercussions of his assessment because of the personal prejudices and the "on-the-record" positions of the Executive Office. And the machinery does not exist where the public can be kept abreast of significant matters concerning U.S.-U.S.S.R. strategic developments without the administration in power attempting to censor and control information to its political advantage; in the past, this information has been leaked to the public via persons affiliated with various U.S. government organizations. I hope that this book contributes to the public data base and the reader has a better understanding of the Soviet political-espionage-academic-military-industrial complex. More importantly, I hope that the reader has a better appreciation of the competence, magnitude, and implications of the Soviet space effort, and their overall strategic forces. If the current trends continue, there is no question that the Soviets will achieve absolute military, strategic, and space superiority over the United States, and should they do so, the Communist Party of the Soviet Union will be among the first to tell us.

What can we do? Let us recognize that we need a massive reorganization of the way we do things, and this means everyone. We are currently in a vicious circle: defense planners claim that they are not getting the needed public support to adequately plan their programs in advance, so that they can reduce costs in the long run; the public is reluctant to support certain programs because of the waste in these establishments. Both groups blame each other, Congress vacillates between what the public wants and what the space and defense establishment requires, and the end result is costly aerospace systems that are frequently poorly planned and must be replaced in the immediate future with still other poorly planned systems. And, in the midst of this, the taxpayer is not given credit for having the intelligence to respond responsibly if he or she is told the truth about the Soviet strategic threat.

The public must assume the main responsibility for our current predicament and the state of our space effort. Yes, it can be justifiably argued that there has been waste in our space program, and poor decisions have been made by high-level officials. I have tried to surface some of these problem areas, but when the question is reduced to its very basics the fact remains that the planners of our space program are the American people, for it is the people who elect the congressmen and senators who determine our space budget and priorities. In the Soviet Union these decisions are made by the ruling members of the Politburo. In the United States we are privileged to live in a democracy, and we therefore have the privilege of determining how our resources will be spent. But with this privilege is a responsibility that we all must share—preserving the national security and the defense of the United States. For if we fail in this task, we fail to preserve our individual freedoms. This is not only the responsibility of officials within the Department of Defense, National Aeronautics and Space Administration, or other agencies. It is the responsibility of our government, and lest we forget, *we* are the government. It is the public who authorizes the programs that these agencies must undertake.

Therefore, the question is: are the people of the United States capable of planning their space and defense programs better than the Politburo of the Communist Party of the Soviet Union? The information presented in the preceding chapters suggests that the answer is no, because the Soviets are very definitely embarked on aggressive aerospace programs that in many ways excell that of the United States in both quantity and quality. But is the answer really no? Or should one consider that the public has not been informed, and therefore cannot logically question or become involved in determining the course of government projects? For all practical purposes, some empire-builders in our government have brainwashed the public into believing that decisions should be left to those in Washington, because that is their job. Some of our problems, but not all, can be traced to self-seeking, empire-building, bureaucratic, incompetent individuals who attained their position because they were either appointed—a political payoff —or were elected by an uninformed or disinterested public. In some cases these incompetent individuals *are* the people in Washington, and the last thing that we need is to leave critical problems for these individuals to solve, because if we do things will get worse and the taxpayer will pay more taxes to support even more poorly planned programs. This problem, by the way, is not only confined to the space and defense industry; in the areas of health, education, and welfare, it is infinitely worse, but this is a subject that should be treated separately.

The American public must (1) recognize that a problem exists, (2) be willing to commit some of its resources to counter the Soviet threat, and (3) demand that Congress responsibly execute its duties. We must permit our best scientists, engineers, and managers to plan on the long-term basis; our projects must be planned in the context of a total national space and defense program, and not as a separate entity. If projects are managed in this manner they will cost the taxpayer considerably less over the long run. Just as a housewife plans the family budget and knows how much money she needs to feed her family, aerospace administrators prepare weekly, monthly, and yearly budgets, and procure aerospace hardware in much the same manner. However, if the housewife's husband is a gambler, and he chooses to gamble with the family's grocery money, she can no longer plan ahead for that Sunday dinner until she has the grocery money in her hand. And if her husband gives her more than she needs on Sunday, when the food stores in her town are closed, that doesn't help either. What she needs is a reliable breadwinner. This is essentially the problem that plagues America today; long-range programs are planned on a yearly basis because NASA and DOD officials cannot be given a guarantee that their programs will be supported throughout their duration. Congress evaluates programs each year, and this practice should be continued, but during election years political considerations have more weight with politicians than good planning, and the budgets of some programs are turned on and off like a water faucet, with the waste in added expenditures passed off to an uninformed public who has accepted that budgetary and administrative matters should be handled by the politicians and administrators in Washington. This is clearly not the way to do things, and the solution to the problem reverts back to the public.

As a starting point, let us recognize that since the peak years of the Apollo Program the National Aeronautics and Space Administration's annual budget is at its all-time low—three billion dollars—compared to an annual United States government budget of about 270 billion dollars. In other words, our space program represents about 1 percent of the total annual federal government expenses. By comparison, about 125 billion dollars, or 47 percent of our annual budget, is expended on human resources, such as health, education, welfare, and public assistance programs; 80 billion dollars, or 30 percent of our annual budget, is allocated toward national defense. Therefore, while there is surely waste in all programs, and I have cited a few examples in our space effort, let us recognize that the real place to look for waste, if we are to relieve our tax burden, is not in our space program but in government human resources and defense programs which spend more than 205 billion dollars annually, or about 77 percent of the total annual federal budget. With these statistics, one can question the competence of the biased commentators and news reporters who have successfully made our space program the scapegoat for the ills of our nation. These are the same men who consistently talk about other priorities. And one should also question the leadership of those space officials who sat on the sidelines, while members of the anti-space establishment tore the space industry apart with untruths and political blackmail. Everyone has heard the expression: "If we can send men to the moon, then why can't we ... ?" Let us remember: NASA sent men to the moon at only a fraction of the cost allocated for health, education, welfare, public assistance, and defense, and this was a feat that we should all be proud of. The engineers who built the hardware to send men to the moon are not responsible for the ill-fated and poorly conceived social programs that currently plague our country.

Plate 31, in my opinion, summarizes the plight of the U.S. space program. Since 1966, NASA's budget has steadily decreased, while national expenditures in the social field have skyrocketed to the point where the taxpayer has a legitimate reason to complain. Yet, some government officials still believe that NASA's budget should be reduced even more, to provide funds for the already poorly managed social programs. This logic is equivalent to the man who beats his wife for spending two pennies more on a loaf of bread, after he lost several thousand dollars at the race track.

NASA is responsible for managing the U.S. space program and those systems such as the space shuttle which will play a major role in the national defense of the United States; let there be no mistake—the new arena for conflict between the superpowers will be near-earth space. Unfortunately for the U.S. cause, NASA has deferred its plans to develop an orbit-to-orbit shuttle and orbital space stations—two of the three space systems that are an inseparable part of a total national space and defense program—because of budgetary reasons. If serious work on these programs is not initiated soon we will be on the verge of another Sputnik era, and possibly a technological Pearl Harbor.

In January 1973, President Nixon exerted pressure on NASA to

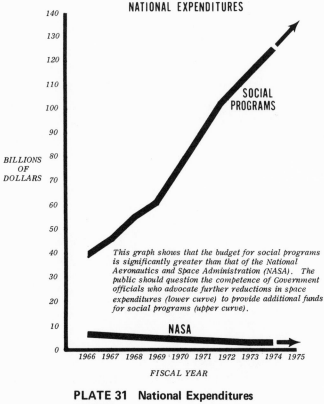

NATIONAL EXPENDITURES

SOCIAL PROGRAMS

BILLIONS OF DOLLARS

This graph shows that the budget for social programs is significantly greater than that of the National Aeronautics and Space Administration (NASA). The public should question the competence of Government officials who advocate further reductions in space expenditures (lower curve) to provide additional funds for social programs (upper curve).

NASA

1966 1967 1968 1969 1970 1971 1972 1973 1974 1975

FISCAL YEAR

PLATE 31 National Expenditures
SAS-46

further reduce its budget, and NASA reciprocated by canceling several projects and announcing that work on the U.S. space shuttle program would proceed at a slower pace. Additionally, NASA has revived its studies on using a smaller payload bay for the space shuttle's orbiter stage because some people believe that the space tug (i.e., orbit-to-orbit shuttle) will not be built. Are we heading in the right direction or are things getting worse?

While NASA's budget is currently about three billion dollars per year, it needs about five billion dollars annually to do an effective job. If the public is willing to assume this financial responsibility (i.e., 2 percent of our total national budget should be allocated to space), then NASA's scientists and engineers must assume their share also. And this means placing emphasis on getting the job done at the least possible cost. We must stop trying to design the most sophisticated hardware, especially when sophistication is unnecessary: planning a program is infinitely more important than advanced technology, though both are important. I offer the crude Soviet standard launch vehicle as an example of what can be done with minimum resources and excellent foresight. The Soviet space shuttle program and its complementary space systems are other examples.

It is time that the U.S. aerospace industry told the public the truth about the causes for some of the waste in the industry. The logic: "What the hell, it's government money anyway," prevails in the industry today, though not to the extent it once did during the 1960s. Additionally, we cannot produce low-cost weapons systems if government officials insist on overspecifying design requirements to justify their own jobs, or the existence of a particular department or agency. We have too many people determining the design characteristics of hundreds of components of certain weapons systems, and these people, to impress their superiors, insist that their components be designed with the most advanced (and costly) materials and with such perfection that the cost of the total system suddenly skyrockets above original estimates. Congress must identify those government organizations that create work and spend money to keep their personnel busy, when their effort is not even remotely related or coordinated with the national effort.

Another wasteful practice is when tax dollars are invested to train engineers who over the years acquire scientific and technological skills and knowledge that are vital to our national defense, and then are laid off because of sudden budgetary problems. A related situation, almost as wasteful, is when experienced engineers from one corporation are layed off and hired by another corporation. The layed off engineers must readjust to their new working environment and often must be retrained at the taxpayer's expense. At a later date, when the contract work at this corporation is completed, these engineers are layed off again. When they are rehired by still another corporation, they are usually retrained, again at the taxpayer's expense, and the cycle goes on, and on, and on. This massive turnover in personnel is not only costly but it prevents defense contractors from forming experienced design teams; this shows up as costly noncompetitive weapons systems relative to the Russian effort.

The public must exert pressure on Congress to undertake measures to minimize the turnover of experienced scientific and technical personnel without violating a worker's constitutional rights; it does not make sense to invest billions of dollars of government money to develop a capability, and then, because of poor planning, have this capability disseminated among the ranks of unemployed engineers who are forced to drive cabs, work in gasoline stations, restaurants, and other nontechnical jobs to earn a living.

To reduce costs and improve the quality of weapons systems, we should consider the creation of a national defense team, to be composed of engineers and managers throughout the country. The idea would be the following: it is in our national interest to develop a capability and to retain it. A small fixed percentage of engineers from each major corporation—the cream of the crop—would be selected by their management to belong to the national defense team, and their salaries would be paid by the government. If the corporation encounters budgetary problems at a later date and must lay off personnel, members of the defense team are untouchable. Thus, design teams in the corporation would stay together. For example, General Electric and Pratt & Whitney Aircraft compete against each other in the jet engine field. In 1967, when General Electric was awarded the supersonic transport (SST) engine contract, Pratt & Whitney was forced to lay off some very good jet engine design engineers. Likewise, when Pratt & Whitney was awarded the F-15 air superiority fighter engine contract by the Air Force in 1970, General Electric was forced to lay off some of its best jet engine design engineers. In both cases, experienced design teams were broken up. However, if the core of Pratt & Whitney and General Electric design engineers belonged to a national defense team, the key design teams would have been left intact after the layoffs. Furthermore, the losing design teams, because their salary is paid by the government, could work on other government programs and advance that technology that would be in our national interest. Thus, when the next engine competition is held, the best design teams in the country would compete against each other, and because the corporation that wins the competition gets to produce and sell the weapons systems developed, the free enterprise profit motive would prevail. A final note on this subject: our country spends billions of dollars on defense, but because of political and economic reasons this money is distributed to numerous corporations throughout the country to spread the wealth, so to speak. The claim is made that the most qualified contractor is awarded the contract, though in reality this is not always the case. A national defense team would not eliminate the award of a contract to an undeserving corporation, but it would prevent this unethical practice from breaking up experienced teams.

The public should recognize that significant corrective action will not be initiated by the government or private industry on their own accord, because they do not have to—it is the public's money that is currently being spent, and if the public does not assert itself these changes will never be made. The public must demand that we be provided with a viable national space and defense program at the lowest possible cost. Concerned citizens must support those congressmen and senators who will take *both* positions (i.e., we need a strong military-industrial complex to protect our national security, and we need to undertake new measures to reduce costs). We are faced with an arduous task: *if we do not strengthen our offensive and defensive strategic forces, we face annihilation in the years ahead; if we do not reduce the cost of deploying new weapons systems, we face bankruptcy.*

Just as it is necessary to minimize the turnover in scientific and technical personnel, we must select competent managers who can work at their position for ten, fifteen, and even twenty years. It is impossible to plan on a long-term basis and thus lower costs and improve the quality of our weapons systems if we shuffle managers on a routine basis. Whereas it is common for Russian managers to hold the same position for tens of years, high-level U.S. government managers appointed by the President are usually shuffled every four years to coincide with the elections, and sometimes more often. For example, since 1968 the United States has had five secretaries of defense (McNamara, Clifford, Laird, Richardson, and Schlesinger) and the National Aeronautics and Space Administration has been headed by four men (Webb, Paine, Low, and Fletcher).

If we are to have continuity we must select the best men in our nation and give them long-term responsibilities commensurate with their capabilities. For example, it would be in our national interest to select someone such as Dr. Wernher von Braun to head NASA and manage our space program for the next ten years or so. And for those readers who feel that Dr. von Braun, who was born in 1912, is too old, it should be recalled that Andrei Tupolev, the designer of the Soviet TU-144 supersonic transport and BACKFIRE supersonic long-range bomber, headed the Tupolev Design Bureau until his death in December 1972; he was 84 years old. In return for assuming this commitment we should guarantee the director of NASA an annual budget on the order of five billion dollars per year for the next ten years. *Given this guarantee, it would be possible for our space planners to design aerospace systems for use on a long-term basis without having to compromise the design of the systems to suit the whims of politicians who have other axes to grind; the taxpayer would get more for his money, and more importantly, he would know that the best his country has to offer would be committed to the space program.* This type of change can only be initiated by people who are tired of paying hard-earned tax dollars only to watch the Soviet Union increase its space lead over the United States because their Politburo plans better than we do.

And finally, our NATO allies very definitely can assist the United States in exploiting space. For example, the West German firm, Messerschmitt-Bolkow-Blohm, is quite knowledgeable about oxygen-hydrogen high chamber pressure rocket engine technology—in 1968 they successfully tested their advanced rocket engine nozzles in Nevada in a joint program with Rocketdyne. It would be in the taxpayer's interest to demand that some of the European firms be permitted to participate in significant post-Apollo programs. Despite the rhetoric, the American taxpayer should know that the Germans in particular would be willing to subsidize their participation in a joint effort with the United States, but because of political reasons their offers have not been accepted. And this is occurring when numerous programs in the United States have been deferred because of budgetary problems. We must think ten and twenty years into the future, and not permit certain individuals with personal axes to grind to interfere with sound long-range planning that is in our national interest.

In the paragraphs below I have summarized the programs that must be supported by Congress and the public to protect our national security.

(1) *The Space Shuttle:* This is the NASA reusable earth-to-orbit launch vehicle that is being designed to deliver up to 65,000 pounds of

payload into orbit per launch. It is vitally needed to support Department of Defense space operations, though publicly attempts are made to justify it on an economic basis for the benefit of mankind. The Russian shuttle is being designed to deliver between 75,000 and 100,000 pounds of payload into orbit per launch, and it will be used to support a far more aggressive space and military program than that in the United States. The Russians lead the United States in the development of a space shuttle by about three years.

(2) *The Space Tug:* This is the reusable orbit-to-orbit shuttle spacecraft that has been studied by the USAF, and recently NASA. However, due to budgetary reasons, its development has been deferred. The space tug (the orbit-to-orbit shuttle) is needed by the Department of Defense to conduct orbital space operations that are beyond the capability of the space shuttle. The Soviets, recognizing the need for orbital spacecraft systems, gained valuable flight experience with the manned Soyuz spacecraft program, and are currently involved in the development of a reusable orbit-to-orbit shuttle that can be carried into orbit by the Rocketoplan—the orbiter stage of their space shuttle. Because the design of the Soviet orbit-to-orbit shuttle is based on modular spacecraft construction—the number of propellant tanks can be varied to accommodate specific missions—and extensive refueling space operations, it will be an extremely versatile space system and an indispensible part of the Soviet military space effort. The Russians lead the United States in the development of orbit-to-orbit spacecraft by about five years, but more importantly, they have plans for expanding this effort while the U.S. does not.

(3) *Space Stations:* Plans to develop orbital space stations by both the U.S. Air Force and NASA have been deferred because of budgetary reasons; the United States has no concrete space station plans after the 1973 NASA Skylab missions; the U.S. news media continues to challenge the merits of Skylab, thereby further deflating public interest in space. The Soviets, recognizing the need for earth orbiting space stations to (1) support their manned orbital spacecraft systems, (2) house scientists for conducting extensive orbital space research, and (3) house orbital weapons systems, gained experience with the Salyut space station launched in 1971. The Soviets plan to launch a 300,000 pound orbital space station in the middle 1970s, and will use their space shuttle to launch orbital space station modules during the late 1970s and 1980s. These modules will be erected by manned and automatic means to form a network of medium size and large orbital space stations. The Russians enjoy about a three year lead

over the United States in the development of space stations and they have plans for expanding this effort while the U.S. does not.

(4) *Long-Range Supersonic Bomber:* This is the Rockwell International Corporation (formerly North American Rockwell Corporation) B-1 bomber that is being developed for the Air Force. The B-1 bomber is urgently needed to replace the aging Strategic Air Command B-52 bomber force of the Korean War era. The Soviets have already developed and flown their own long-range supersonic bomber, which carries the NATO code name BACKFIRE. BACKFIRE can travel about 1,500 miles per hour, can reach strategic targets in the continental United States, and is currently operational. The B-1 will not become operational until 1978–1980. The Russians enjoy about a five year lead over the United States in the deployment of supersonic bomber aircraft.

(5) *Air Superiority Fighter:* This is the McDonnell Douglas Corporation F-15 air superiority fighter being developed for the Air Force. The F-15 represents the first major fighter aircraft development program in the United States in about two decades. World War II, the Korean War, and the Israeli Six-Day War demonstrated the importance of air superiority in fighting conventional wars. The most advanced operational fighter aircraft in the world is the Russian MiG 25, NATO code name FOXBAT, which was first flown in the middle 1960s and can fly better than 2,000 miles per hour. The F-15 will provide the United States with air superiority at low altitudes, while FOXBAT will offer the Russians air superiority at high altitudes. (These aircraft were designed to operate at their maximum effectiveness at different altitudes.) The Soviets are currently developing an air superiority fighter that will be deployed in the near future that will outperform the F-15 in its own low altitude operational environment; the KGB has played a significant role in providing Soviet aircraft designers with details about the operating characteristics of the F-15. The Russians enjoy at least a three to five year lead over the United States in the deployment of air superiority fighter aircraft.

(6) *Undersea Long-Range Missile Systems:* The Navy's Trident submarine program is needed to provide the United States with a first-rate submarine launched ballistic missile (SLBM) strategic capability for the late 1970s and 1980s. The Trident is needed to replace outdated U.S. submarines; the terms of the SALT agreement permit the Soviets to build more strategic submarines that can carry more nuclear intercontinental ballistic missiles than the United States. Reliable information indicates that the Soviets will enlarge their

undersea nuclear strike capability over the next five years, and when combined with their planned land and space-based forces, the strategic balance will be tilted heavily in the Soviets' favor. The Russian Navy is second to none, and this situation is not expected to change in the foreseeable future.

The survival of the United States as a free nation is dependent on our will to support those programs that are critical to our national security. And this includes our determination to support research and development programs that upgrade our knowledge of scientific and technological matters that have a long-range bearing on our national security. The Soviets are extensively involved in defense-related research and development programs. If the current trends continue, the quality of Soviet weapons by 1980 could exceed those of the United States, and the SALT agreements have already assured the Soviets superiority in numbers of offensive strategic weapons systems. The American public's current anti-space, anti-defense position is sometimes understandable, but coupled with numerous unsound political decisions made in recent years, the United States is vulnerable to a Soviet technological Pearl Harbor—the development of a new weapon system that could drastically shift the balance of power overnight. This could come to fruition by the 1980s unless we change our way of doing things and keep abreast of the Soviet effort. If the Soviets achieve a technological breakthrough in weapons systems, or are permitted to further improve their strategic position relative to the United States, our country will cease to exist as a free nation; we would be ruled by an elite group of men in the Kremlin, known as the Politburo of the Central Committee of the Communist Party of the Soviet Union, and we would fare no better than Hungary, Czechoslovakia, and the other Eastern bloc nations under Soviet rule. It is time that Americans recognize that the Soviet government believes that war is a tool of politics, and negotiations are tools for enhancing the Soviet strategic position relative to the rest of the world. Given the proper conditions, the Soviets believe that they can win a nuclear war, and this partially explains their tremendous overkill nuclear arsenal. And let it be known that if there were a nuclear war today, the United States would be completely annihilated, whereas the possibility exists that the Soviets would be only partially destroyed, due to their first-rate defense network and civil defense program. Some hardline Communist strategists would be willing to settle for this today, because the world would be theirs to remake in their own image. And finally, we should recognize that our government *can* be overthrown; a secret surrender is a near-term reality, made possible because of advances in weaponry and an era of secret negotiations. Under conditions of a secret surrender, all U.S. armed forces would be under the control of the President and Department of Defense, and if our leaders are convinced that a checkmate position truly exists, they will do everything possible to avert a national catastrophe—the annihilation of the United States by the Soviet Union. With the control of our military forces in the hands of the very people who would surrender to the Russians, if history is any guide for things to come, then U.S. military units would be used to keep dissident forces and right wing groups from resisting a Soviet takeover; martial law would be declared and tanks would be paraded up and down our streets until the reins of our government have been securely transferred to the Kremlin and Russian troops have gained control of the situation. The public must recognize that by 1980 it will be possible for the Russians to win a war without firing a single missile, because they have created a weapon, or have deployed their strategic land, sea and space-based offensive and defensive forces in such a manner that the United States would be annihilated if the President attempted to resist the General Secretary's demands. The Soviets are heading for this overwhelming superior position, and should this happen, a secret surrender would be inevitable.

I have done my best to reveal in layman's language information about the Soviet space and defense establishment that rightfully belongs in the public domain, but I have only scratched the surface. Thousands of pages of intelligence reports and private memorandums bear me out on this. In the past, the destiny of our nation has been in the hands of empire-builders and leaders who have preferred to look after themselves, or suit their personal prejudices, rather than act in our national interest. Some of these men, by their reports and recommendations to and from the Executive Office, have adversely contributed to the quality of U.S. strategic forces and the foreign policy of the United States. The significance of the Soviet strategic threat in the years ahead also raises questions about the character, competence, dedication, and qualifications of those men who aspire for the office of the President of the United States, and answers why some politicians, who are more popular than competent, should never be elected to this office, whether in 1976, 1980, or any other year.

And finally, a comment to the reader who believes our country is invincible despite Soviet weapons systems advancements. I have often heard the following rebuttal from well-meaning, yet uninformed persons: "Well, if the Soviets are doing it, we must be too ... No one really knows what secret programs 'they' are working on in this country." I would suggest that you invite "them" to dinner and ask about the significant secret programs that our country is supposedly working on; you would be shocked to learn what we are not doing, compared to what you thought we were doing. And be sure to ask about the well-polished government machine (that does not exist), or about the master plan and all-knowing individuals in our government who supposedly perform their daily functions with the flawless efficiency of the government men depicted on television. Whether we like it or not, we live in a real world, with real men; their is no Santa Claus and there are no government men who can offer a rational reason for every move our government makes. Do not assume that "they" are looking after our national security, because "they" exist in your imagination only. The destiny of our nation belongs in your hands, and those persons who you elect to higher office. The chess match is truly between the Politburo of the Central Committee of the Communist Party and the American people, and the next move is yours.

Appendix I

Excerpts from my August 1969 memorandum to the CIA: The memorandum was classified by the CIA to protect me from possible recriminations from the Eastern bloc. The excerpts are presented here as they would appear to a U.S. government intelligence analyst; the format is similar to computerized reports that are disseminated throughout the U.S. intelligence community.

NO FOREIGN DISSEMINATION

EXCERPTS FROM REPORT

REQUIREMENTS LIST: This package is forwarded for your records and/or future use. It outlines the general questions that I will attempt to have answered during the conference and contains the names of some scientists who have attended past IAF Congresses. ...Should you wish to use these questions during the XXth IAF, please be sure to brief prospective U.S. scientists attending the conference that the questions are indicative of what is required, and should never be posed to the individual as it is presented within. The answers can be ascertained through numerous questions and discussions of their own choosing. In the past a number of U.S. scientists have compromised themselves by asking the same individual the same remote question.

███████████ What is the status of their (USSR) plasma engine program? Are plasma engines being considered for orbital delta velocity as well as for attitude control of spacecraft? Are plasma engines being considered for maneuverable satellites? What is the status of their nuclear engine program? Is nuclear propulsion being considered for a lunar shuttle vehicle, as well as for interplanetary missions?

███████████ Get information on the payload requirements of future Soviet space missions; relate this to the Proton and "Super Booster" launch vehicles...remarks on MOL versus Moon versus Mars type programs.

███████████ ...What is the Soviet's position on cooperation in space? Would they be interested in an agreement with the U.S. along the guidelines of the current French-Soviet space cooperation pact?...Is he involved with the Soviet space program anymore?

Duboschine, Georgiy: ...Are Professor G. L. Grodzovskiy's low thrust trajectory studies

<table>
<tr><td>THIS DOCUMENT CONTAINS INFORMATION AFFECTING THE NATIONAL DEFENSE OF THE UNITED STATES WITHIN THE MEANING OF THE ESPIONAGE LAWS, TITLE 18 U. S. C. SECTIONS 793 AND 794. ITS TRANSMISSION OR THE REVELATION OF ITS CONTENTS IN ANY MANNER TO AN UNAUTHORIZED PERSON IS PROHIBITED BY LAW</td><td>1/4
SECRET</td><td>**NO FOREIGN DISSEMINATION**
GROUP 1
EXCLUDED FROM AUTOMATIC DOWNGRADING AND DECLASSIFICATION</td></tr>
</table>

maneuverable satellite-related or interplanetary-related?

███████████ Find out more on their ███████████ Is this effort funded by the Soviet Union?

Neumann, Karl (East Germany):...What is his connection with Soviet Research teams?... He has implied that the Soviets will refuel their spacecraft in space. Was he speculating or does he know? If he knows, who told him?

Pfaffe, Herbert (East Germany): ...What is his connection with the Soviet Research teams?...In the past he has stated that Soviet space stations would serve both as launching platforms for lunar and interplanetary missions. Who told him?

Vinograd, B. V.: ...How many more Venus-type probes have the Soviets planned? When?

Maykapar, G.: Get more information on his work on ███████████ for lift-reentry vehicles...Is he testing at the Central Aerohydrodynamics Institute...Obtain his exact relationship with Professor G. L. Grodzovskiy, O. M. Belotserkovskiy, and G. G. Cherniy.

Belotserkovskiy, Oleg M.: ...Get the latest status on his reentry studies using 3 dimentional analysis...Get his relationship with Professor G. L. Grodzovskiy, G. I. Maykapar, and G. G. Cherniy.

Cherniy, G. G.: ...His institute (Note: Scientific Research Institute of Mechanics, Moscow State University) appears to be most certainly involved in research relating to ███████████ What is the status of this effort? How many people are involved? Who?...Is he conducting tests for Maykapar's lifting body?...Is he coordinating his supersonic combustion studies with G. Petrov's Institute of Space Research?...How about their work on a H_2 Scramjet?...Has he tested...above Mach 12...Does he know Professor G. L. Grodzovskiy? What is their relationship?

<table>
<tr><td>THIS DOCUMENT CONTAINS INFORMATION AFFECTING THE NATIONAL DEFENSE OF THE UNITED STATES WITHIN THE MEANING OF THE ESPIONAGE LAWS, TITLE 18 U. S. C. SECTIONS 793 AND 794. ITS TRANSMISSION OR THE REVELATION OF ITS CONTENTS IN ANY MANNER TO AN UNAUTHORIZED PERSON IS PROHIBITED BY LAW</td><td>2/4
SECRET</td><td>**NO FOREIGN DISSEMINATION**
GROUP 1
EXCLUDED FROM AUTOMATIC DOWNGRADING AND DECLASSIFICATION</td></tr>
</table>

Moisseev, Nikita: Find out if his Moscow Computer Center has frequent contact with the Moscow Physical Technical Institute (MPTI) and the Central Aerohydrodynamics Institute (TsAGI.)

Grodzovskiy, G. L.: ...He could be the key coordinator or primary consultant on a Soviet boostglide vehicle (Rocketoplan) development program. He is very close to the people in the various disciplines who would be organized to undertake such a program. Find out as much as possible...Important - What is his relationship with G. Maykapar, Oleg M. Belotserkovskiy, G. G. Cherniy, Vsevolod Borisenko, and Nikita Moisseev?...Does he work or consult at the Central Aerohydrodynamics Institute (TsAGI) in addition to his work at the Moscow Physical Technical Institute (MPTI)...Does he personally know designer Mikoyan?...Get his views on composite - rocket/airbreathing engines...Find out if large plasma engines are being worked on in the USSR for delivering orbital delta velocity... What are his opinions on the multichamber plug cluster-aerospike engine...Will the Soviets use this type of propulsion system in the future?

PRELIMINARY ANALYSIS: This package contains information on Soviet scientists and facilities that could be involved in a boostglide vehicle (Rocketoplan) development program...The activities of the following engineers and scientists should be watched closely: (1) G. L. Grodzovskiy; (2) G. I. Maykapar; (3) O. M. Belotserkovskiy; (4) Nikita Moisseev; (5) G. G. Cherniy; (6) V. M. Borisenko. It appears that these individuals are involved in a classified development program concerned with lift reentry vehicles. We have called these boostglide vehicles. The Russians have referred to them as "Rocketo-plans." Soviet open literature sources have stated that the Soviets will build a Rocketoplan. I have outlined below some fragmentary information that links these people together on such a program: (1) G. Maykapar is affiliated with the Moscow Physical

3/4

SECRET

GROUP 1
EXCLUDED FROM AUTOMATIC DOWNGRADING AND DECLASSIFICATION

EXCLUDED FROM AUTOMATIC DOWNGRADING AND DECLASSIFICATION

SECRET NO FOREIGN DISSEMINATION

Technical Institute...recently (Oct. 1968) presented a paper, "Aerodynamic Heating of the Lifting Bodies"...; (2) O. M. Belotserkovskiy is director of the Moscow Physical Technical Institute...He is doing theoretical work on lifting bodies. During the past, Belotserkovskiy was director of the Computer Center of the Academy of Sciences...; (3) N. N. Moisseev ███████████████████████ at the Computer Center of the Academy of Sciences and could be the head of the Center. He is a mathematician and aerohydro-dynamicist...; (4) G. G. Cherniy is the director of the Scientific Research Institute of Mechanics at Moscow State University. He has recently ████████████████████ on lifting bodies; (5) V. M. Borisenko is a professor of mathematics...Moscow State University...██████████████ is certainly applicable to a Soviet Rocketoplan; (6) G. L. Grodzovskiy is listed as a professor at the Moscow Physical-Technical Institute and is a physicist, mathematician, and aerohydrodynamicist. He was affiliated with the Central Aerohydrodynamics Institute imeni N.Ye. Zhukovskiy (TsAGI)...and is probably a consultant at this time. He may have worked for G. I. Petrov during the 1950's... Grodzovskiy is very knowledgeable of the shortened nozzle concept...and was seen in the company of Borisenko during a 1966 international conference (Note: Madrid IAF Congress). During the middle 1960's he showed considerable interest in reusable spacecraft. He is a friend of Moisseev and Cherniy, as demonstrated at various conferences and appears to be quite knowledgeable of systems integration and design techniques. He recently authored a paper, "On Bodies of Revolution Having a Minimum Total Drag Coefficient and Low Heat Transfer at High Supersonic Flight Speeds," which is similar to Cherniy and Maykapar's area of interest. He is close to Belotserkovskiy on a personal basis. I feel that G. L. Grodzovskiy is the primary consultant in this program, and he is very knowledgeable in the aforementioned areas of study and is very acquainted with the specialists working in these areas...In summary, the following facilities should be watched closely for work relating to a Soviet Rocketoplan: (1) Central Aerohydrodynamics Institute imeni N.Ye Zhukovskiy (TsAGI); (2) Moscow Physical Technical Institute (MPTI); (3) Scientific Research Institute of Mechanics at Moscow State University; (4) Computer Center of the Academy of Sciences, Moscow.   4/4

Appendix II

Excerpts from my Paris and Mar del Plata reports to the CIA: These excerpts were transcribed from my personal notes and are presented here as they would appear to U.S. government intelligence analysts. The word "source" refers to me. (Note: raw intelligence reports are usually classified "Confidential—No Foreign Dissemination" or "Secret—No Foreign Dissemination" by the CIA and the DIA.) For the examples shown, the marking "No Foreign Dissemination" means that foreign nationals, such as European officers from NATO, would not be permitted to have access to the report; only persons with a Department of Defense "secret" clearance and a legitimate reason for using the document—termed "need to know" in security parlance—would be permitted to have access to the report.

TITLE-- Excerpts From Report

AUTHOR-- None

NO FOREIGN DISSEMINATION

COUNTRY OF INFO-- USSR

SOURCE--None

DATE OF INFO--May-June 1969   DATA CATEGORY--Science and Technology

PROFILE HITS--Rocketoplan, Boostglide Vehicle, Propellants, Refueling in Space, Rocket Engines, Boosters

SUBJECT AREAS--Physics, Chemistry, Aerospace Technology

TOPIC TAGS-- R&D Personnel, Rocketoplan, Boostglide Vehicle, Biographical, Propellants, Refueling in Space, Rocket Engines, Budgets, Science City, Jet Engines, Compressors, Nozzles, Boosters, Pumps, Chambers

CONTROL MARKING-- Classified

DOCUMENT CLASS-- Secret

PROXY REEL/FRAME--9361/9721   INTELL REPORT NO-- 9814638191

CIRC ASSESSION NO-- P90651319

USER-- C11, C13

1/6

S E C R E T

GROUP 1
EXCLUDED FROM AUTOMATIC DOWNGRADING AND DECLASSIFICATION

S E C R E T

This report contains information obtained by source while attending the 1969 Paris Air Show during May-June 1969. This report is classified Secret - No Foreign Dissemination to protect source and his access to information. Organization and evaluation of this and other raw intelligence information will be incorporated into source's annual report early next year.

EXCERPTS FROM REPORT

ROCKETOPLAN - BOOSTGLIDE VEHICLE:

███ was asked by the source if the Soviets were working on a program similar to the US HL-10 Program (source showed ███ a photograph of the HL-10 lifting body that was shown on page 18 of an official air show brochure, "Information and Documents Numero Special Salon du Bourget".) ███ sketched a schematic of the Earth and a manned orbital space station. He stated that the Soviet "Rocketoplan" would be used to transfer men between Earth and the space station. He stated that the orbital space station would be supplemented by a "Rocketoplan" with "many reuses." He said that the Soviets have considered three propellant combinations for use in the "Rocketo-plan": (1) Nitric Acid - Kerosene (2) Oxygen-Hydrogen; (3) Fluorine-Hydrogen. When the source sketched a lifting body, ███ wrote the chemical formulas of oxygen-hydrogen and fluorine-hydrogen under the sketch. ███ used the terminology "boost assist" and "Rocketoplan" in discussing the boostglide vehicle. When asked about the status of the system, ███ stated that the Soviet "Rocketoplan" has aerodynamic control in the atmosphere and that an unmanned-unpowered version of the Soviet "Rocketo-plan" had already been tested.

NO FOREIGN DISSEMINATION

2/6

S E C R E T

GROUP 1
EXCLUDED FROM AUTOMATIC DOWNGRADING AND DECLASSIFICATION

REFUELING IN SPACE: Source asked ███████████████████████████
███████if the Soviets intended to refuel spacecraft in near-Earth space. ████████
stated that the "transfer of fuel (propellant) in space has already been confirmed."
He stated that Kosmos 186 and 188 and Kosmos 212 and 213 (automatic rendezvous experiments) had demonstrated the feasibility of electrical and mechanical linkage. Source
drew a sketch of two spacecraft and asked ████████ if the propellant transfer would be
conducted with a flexible propellant transfer line. ████████ said "no!"and drew an
"X" over the propellant transfer line that source had drawn. ████████ then drew a
schematic similar to the one shown below:

MEN   I   AUTOMATIC

SIMILAR TO SCHEMATIC DRAWN BY
████████ USING HIS NOTATION   II

He said that the transfer of propellant would be conducted using direct mechanical
linkage and the difference in tank pressure between the spacecraft. He stated that
spacecraft number II was primarily a propellant supply spacecraft and would not have
a capsule (command module.) The transfer of propellant would be conducted automatically.
The Soviets considered using cosmonauts to assist in transferring propellants and concluded that the automatic mode was more efficient and safer. Source asked ████████ if

any propellants had been transferred in space to date. ████████ stated that nitric
acid and kerosene have already been transferred successfully in space. Source asked
if other propellants were being considered. ████████ stated that oxygen-hydrogen was
a candidate fuel (propellant) because of its high specific impulse. He also stated
that the transfer of oxygen-hydrogen propellant has been demonstrated successfully by
the Soviets on Earth, but not in space. He indicated that storability was very important
and this factor made the problem very difficult if oxygen-hydrogen were used. He stated
that fluorine-hydrogen was not desirable because of its toxicity. ████████ implied
that both orbital spacecraft and large space stations would make use of the automatic
rendezvous mode for refueling in space.

RD-214 ROCKET ENGINE: ██
stated that the RD-214 rocket engine uses a "special blend" of kerosene for the fuel
and nitric acid for the oxidizer. Source asked ████████ if the Soviets had encountered
any combustion instability with the engine and if either of the propellant components
used any additives. ████████ replied, "no" to both questions and emphasized that no
additives were used. He stated that the oxidizer (nitric acid) was used to cool the
chamber and that kerosene was injected directly into the chamber. (Source Note: This
appears to be an error). Both injectors for the oxidizer and fuel are of the centrifugal swirl type (████████ used the expression "swirl cone spray" for each
injector.) ████████ noted that (1) "many" injectors were used and (2) the spray patterns
of the individual cones mixed in the chamber, providing a good mixture ratio profile
across the injector head. He stated that the coolant passages of the thrust chamber
were helical. He demonstrated the nitric acid flow schematic by rotating his arm in a
spiral fashion, starting near the inlet manifold (about five inches from the exhaust

SECRET segment...

S E C R E T

nozzle exit plane) and ending near the injector-combustion chamber assembly. He noted
that the thrust of the engine was 72 tons and not 74 tons as was shown on the placard
(Each chamber delivers 18 tons thrust.) Source asked███████if the designation RD-214
meant that the engine was designed after the RD-107 and RD-119 engines. ███████ replied
that attention should not be paid to the number designations because many rocket engine
projects, whose engine designations vary greatly, were started in parallel with each
other. ███████ stated that the RD-214 engine is used in the Kosmos carrier rocket,
whose upper stage engine was released earlier (source concluded that███████was
referring to the████Kosmos launch vehicle, which uses the RD-119 rocket engine in its
upper stage.) Source (1) drew a schematic of a cross section of a first stage, containing four main engines in the booster stage and four strap-on modules and (2) asked if the
schematic resembled the Kosmos carrier rocket booster stage. ███████ drew an "X" over
the strap-on modules, stating that strap-ons were not used. Source asked if four main
engines were used on the booster stage. ███████said yes. (Source indicated that
███████could have referred to four (4) chambers rather than four main engines.
███████was interrupted by "embassy type personnel" and source was unable to pursue the
point.)

    OXYGEN-KEROSENE ROCKET ENGINES: ███████████████████████████████
███████and an unidentified associate stated that 120 atmospheres (1760 psia) was the
highest chamber pressure that the Soviets considered for an oxygen-kerosene rocket engine.
    HIGH CHAMBER PRESSURE OXYGEN-HYDROGEN ROCKET ENGINE: ███████████████
███████████████ stated that the high chamber pressure oxygen-hydrogen staged
combustion rocket engine currently being developed for the Air Force was a "very advanced
engine." ███████also referred to the engine as an "ambitious design." Source asked

5/6

S E C R E T

S E C R E T

███████ why the Soviets did not bring an oxygen-hydrogen rocket engine to the exhibit.
███████ replied, "We have many engines, we couldn't bring them all."
    ZOND-6: ███stated that Zond-6
(Zond spacecraft) could be modified to accommodate cosmonauts. He noted that the Zond
reentry module was not the same as the Soyuz reentry module because the Zond system was
designed for a different cosmic velocity.
    LUNAR MISSION: ███ stated that
there were two methods of landing men on the moon. (1) The Saturn V approach and (2)
The Earth-orbit rendezvous of Proton and Soyuz spacecraft (source noted that the use of
Soyuz indicates that the Proton launch vehicle is not yet man-rated and the Soyuz Booster
will probably be used in the early 1970's to support the Soviet near-Earth man in space
program.
    PROTON BOOSTER - PROTON SATELLITE: ██████████████████████████████
███████ stated that the payload of the Proton booster can be increased beyond 17 tons
(the weight of the Proton IV satellite on exhibit.) Source asked if the diameter of the
Proton IV satellite (cylindrical portion, with the solar panels retracted) was identical
to the diameter of the Proton booster's last stage. ███████ replied that it was.

6/6

S E C R E T

SECRET

TITLE-- Excerpts From Report

AUTHOR--None

COUNTRY OF INFO-- USSR

SOURCE-- None

DATE OF INFO-----October 1969 DATA CATEGORY--Science and Technology

PROFILE HITS--Rocketoplan, Boostglide Vehicle, Space Shuttle, Propellants, Materials,
　　　　　Rocket Engines, Boosters, Chambers, Intelligence Operations
SUBJECT AREAS--Physics, Chemistry, Aerospace Technology, Political Science

TOPIC TAGS--R&D Personnel, Rocketoplan, Boostglide Vehicle, Biographical, Propellants,
　　　　　Refueling in Space, Space Stations, Rocket Engines, Budgets, Science City,
　　　　　Jet Engines, Compressors, Nozzles, Boosters, Pumps, Chambers, Intelligence
CONTROL MARKING-- Classified

DOCUMENT CLASS-- Secret
PROXY REEL/FRAME--9763/9881　　　　INTELL REPORT NO--9874513620

CIRC ACCESSION NO--P95610216
USER-- C11, C13

1/32

SECRET

SECRET　　NO FOREIGN DISSEMINATION

his report contains information obtained by source while attending the 20th International
stronautical Congress in Mar del Plata during 3-11 October 1969. This report is
lassified Secret-No Foreign Dissemination to protect source and his access to informa-
ion. Organization and evaluation of this and other raw intelligence information will
e incorporated into source's annual report early next year.

EXCERPTS FROM REPORT

bservations made at the Metropolitan Buenos Aires Airport Friday evening (5:30 p.m.-
:00 p.m.), 3 October 1969; Source arrived at the Metropolitan Buenos Aires Airport and
et Professors Georgiy Zhivotovskiy (spelled Jivotovski in the official list of attendees)
nd Leonid Sedov, who were waiting to depart on flight 640 for Mar del Plata. Prior to
he flight Zhivotovskiy greeted approximately seven (7) "embassy type" individuals.
Source noted that he did not see any of these individuals during the conference and con-
luded that these individuals were concerned with the Soviet intelligence apparatus that
perated near the Hotel Provincial in Mar del Plata.) Source overheard one of the in-
ividuals inform Zhivotovskiy that after the conference, Zhivotovskiy was to contact a
an named "Leoli" (phonetic spelling) who was located in Cordoba and "would take care of
verything."

Discussion with Gennadi Dimentiev on Flight 640 (Buenos Aires to Mar del Plata);
riday Evening (7-9 p.m.) 3 October 1969: BOOSTGLIDE VEHICLE - ROCKETOPLAN - REUSABLE
PACE TRANSPORT SYSTEM: Source learned that Dimentiev is affiliated with the Moscow
viation Institute (MAI) and apparently is a specialist in systems design and analysis.
hen asked whether he was a rocket or airbreathing specialist, Dimentiev indicated that
e was concerned with both fields. In a discussion on boostglide systems (Rocketoplans
s the term used by the Soviets), Dimentiev stated that the recirculation of a rocket

2/32

SECRET

A. G. M. ZHIVOTOVSKY

B. YOURI ZONOV

C. IGOR PRISSEVOK

D. JOULI KHODAREV   YOURI ZONOV   G. DIMENTIEV

E. G. M. ZHIVOTOVSKY
MAR DEL PLATA, ARGENTINA
G. M. ZHIVOTOVSKY
L. BASHCKIROV
JOULI KHODAREV
OCTOBER 3-11, 1969
G. DIMENTIEV

F. IGOR PRISSEVOK   JOULI KHODAREV

G. IGOR PRISSEVOK   YOURI ZONOV

H. HORST HOFFMANN
(EAST GERMANY)

I. EBERHARD HOLLAX
(EAST GERMANY)

engine's exhaust products on the upper half of the Rocketoplan is a very severe problem that the Soviets were concerned with. Dimentiev stated that the Soviets would probably use a fairing or shroud, as indicated below, to augment the thrust of a rocket system while in the atmosphere.

SHROUD

NOZZLE   **CHAMBER**

He indicated that a shortened nozzle contour was attractive for use in the Rocketoplan, also adding, that the recirculation of gases at the base of the Rocketoplan is small if the engine is small relative to the base of the vehicle. Dimentiev asked (1) whether the US intended to recover both stages of their system, (2) would the first stage go into a low earth orbit, and (3) whether the range of the first stage permitted it to fly over the USSR if launched from the US. In addition, if the US were to use a Titan/Rocketoplan configuration, Dimentiev asked source what the stage separation velocity and altitude would be. Source concluded that Dimentiev is quite familiar with Soviet Rocketoplan designs and is probably very much involved in the Soviet Rocketoplan program as a systems design engineer. During the flight to Mar del Plata, source noted that Dimentiev's first name as indicated on his passport was Gennadi. However, his first name initial on his personal briefcase was the Soviet equivalent of the English letter "p" (the Soviet letter "peh", which looks like the Greek letter "pi".) In later sections of the report, source notes that an intentional effort was made to "play down" Dimentiev's activity with Soviet hardware development programs and his importance in these areas.

Discussion with G. Zhivotovskiy Sunday Afternoon (12:30 - 3:00 p.m.) 5 October 1969:

3/32

S E C R E T

GROUP 1
EXCLUDED FROM AUTOMATIC DOWNGRADING AND DECLASSIFICATION

S E C R E T   NO FOREIGN DISSEMINATION

CLANDESTINE INVITATION TO USSR: Professor Zhivotovskiy invited source and "number one" to visit the Soviet Union for 21 days during 1970. He stated that all expenses associated with the visit after arrival and prior to departure from Moscow would be incurred by the USSR Academy of Sciences. Source and "number one" were promised an itinerary that included Moscow, Leningrad, Science City near Novosibirsk, and one more city (possibly Odessa or Sevastopol.) Zhivotovskiy stated that most of the engineers in source's main area of interest were located in the Moscow area. Source was promised a tour of the scientific institutes and an opportunity to meet Soviet engineers to discuss subjects of source's own choosing. Source and Zhivotovskiy agreed to meet in Moscow during August or September 1970. Source was instructed to cable the Professor at the Academy one week prior to the arrival date, so that Zhivotovskiy could make the necessary arrangements. Zhivotovskiy said that he would notify the Soviet Embassies in Paris, Rome, Belgrade, and Athens on 1 August 1970 to prepare visas for source and "number one." Source was instructed to pick up visas at any one of these embassies after 1 August. Zhivotovskiy indicated that he would personally escort source and "number one" on most of the visit, including the trip to Science City. He also indicated that if source planned to attend the 21 IAF Congress in Germany after his visit, there was a possibility that source could fly to Germany with the Soviet delegation. US-USSR COOPERATION: Source asked Zhivotovskiy if the Soviets would be interested in exchanging old and unclassified aerospace data and possibly hardware with the US. Zhivotovskiy indicated that at the Academy he was concerned more with science than with engineering problems. However, he would bring up the question with Professors G. L. Grodzovskiy and G. Maykapar, whom source had mentioned in previous discussions with the Professor. Zhivotovskiy stated that Grodzovskiy and Maykapar were engineers and could discuss this with source in Moscow next

4/32

S E C R E T

GROUP 1
EXCLUDED FROM AUTOMATIC DOWNGRADING AND DECLASSIFICATION

year. Source noted that in the past both Grodzovskiy and Maykapar have referred to themselves as theoreticians and even denied having knowledge of hardware programs.

SOVIET TEXTBOOKS: Source asked Zhivotovskiy about several new Soviet textbooks that were recently released. Zhivotovskiy promised to have these textbooks waiting for source upon source's arrival in Moscow and indicated that source could obtain others while visiting the USSR.

MEETING BETWEEN KOSYGIN AND CHOU EN-LAI: Source asked Zhivotovskiy about the results of the meeting between Premier Kosygin and Chou En-lai. Zhivotovskiy stated that Kosygin's stop was short ("a few hours") and compared it to waiting for a plane at an airport. He stated, "They accomplished nothing." He indicated that Kosygin had wasted his time. When asked if Kosygin felt slighted because Mao Tse-tung or Lin Piao did not meet him, Zhivotovskiy stated that the protocol was in order, as Kosygin and Chou En-lai had comparable positions. He stated that protocol was not as important as "substance" and they "accomplished nothing."

TRANSPORTATION OF SOVIET DELEGATION: Zhivotovskiy stated that most of the Soviet delegation to the conference flew Aeroflot from Moscow to Dakar, with a stop in Algeria for refueling. After staying in Dakar one night, they flew Lufthansa to Buenos Aires.

3RD CONFERENCE ON SPACE ENGINEERING: Zhivotovskiy expressed a desire to influence the selection of the location for the 3rd International Conference on Space Engineering, which was being organized by Dr. Partel for the spring of 1971. Zhivotovskiy thought that Partel's 2nd International Conference on Space Engineering (May 1969 -Venice, Italy) was very worthwhile.

ZHIVOTOVSKIY'S TRAVELS: Zhivotovskiy said that he has visited Japan, The Philippines

5/32

S E C R E T

Europe, Asia, North America, and South America and expressed a desire to visit Australia and the South Pacific Islands.

Observations made during Dr. George Mueller's invited paper during 20 IAF Congress; Monday morning (11:00 a.m. - 11:45 a.m.) 4 October 1969; SOVIET REACTION TO APOLLO 11 FILM: Gennadi Dimentiev, Moscow Aviation Institute, and Professor Jouli Khodarev (pronounced Hodarev; deputy director of the Institute of Space Research, Moscow) became very excited during the NASA Apollo 11 Lunar landing sequence that was shown on film after Dr. Mueller's presentation. Professor Khodarev indicated to Dimentiev that the LEM descent engine had lateral maneuverability while hovering over the touchdown, by moving his hands laterally above the floor. When the LEM touched down both men literally let out a sigh of relief. Their rapid conversation after lunar touchdown indicated to source that the Soviets are especially concerned with the final seconds of the descent trajectory before lunar touchdown. Aside from the lunar landing sequence, the slide from Dr. Mueller's presentation that received the greatest response from Dimentiev and Khodarev showed an artist's conception of US space shuttle systems (reusable upper and lower stage vehicles.) Dimentiev rapidly recorded the significant characteristics of the US space shuttle systems on his note pad. Professor Georgiy Zhivotovskiy commented that he was not impressed with the Apollo film. He said that he was impressed with "the fact" but not in the propaganda. The film showed nothing of technical significance. He also remarked that he was happy that Armstrong, Aldrin, and Collins did not stay for the conference because they would have spoiled the technical flavor of the meeting. Zhivotovskiy said that since the US had beaten the Soviets to the moon, the Soviets were no longer in a hurry to get there. The source noted that earlier in the year, Zhivotovskiy had indicated that there was no moon race. In this discussion, however, he admitted that they were

6/32

S E C R E T

racing the US to the moon and lost.

Technical Discussion with Oleg Belotserkovskiy, Vladimir Sychev, Igor Prissevok, Gennadi Dimentiev, Yuri Zavernaev, and Georgiy Zhivotovskiy Monday Afternoon (12:45 p.m.- 2:45 p.m.) 6 October 1969; REUSABLE LAUNCH VEHICLES - SHORT NOZZLES: Source renewed his short acquaintance with Gennadi Dimentiev and proceeded to continue their discussion on reusable launch vehicles. Dimentiev called in Igor Prissevok to help in the translation. As the discussion progressed, Professor Oleg Belotserkovskiy and Yuri Zavernaev joined in. Source had heard about a paper written by a Soviet engineer named Gogish, who is an internal aerodynamicist and nozzle design specialist, and inquired whether anyone knew him. All stated that they had not heard of Gogish. Source continued by sketching a rocket nozzle contour as shown below:

Source showed that the Gogish shortened nozzle contour had a relatively sharp expansion downstream of the throat, relative to US nozzles, but was considerably shorter than US nozzles. The Soviets did not respond. The source continued by sketching a lifting body and stating that US studies show that a Rocketoplan should contain a high performing rocket propulsion system and that it is desirable to use a short nozzle. Therefore, the Gogish nozzle would be very attractive for a Rocketoplan application. When source referred to Maykapar's paper at the 19th IAF Congress in New York, which contained a

7/32

S E C R E T

theoretical discussion on lifting bodies, Belotserkovskiy rudely interrupted source, yelling, "Maykapar works at my institute (Moscow Physical Technical Institute)...Why do you associate Maykapar with Gogish. Who is this Gogish? Why do you associate Maykapar with a Rocketoplan?" Source is certain that he hit a sore spot, and Belotserkovskiy reacted instinctively. Belotserkovskiy's yelling brought in Professor Zhivotovskiy, who heard him from the other side of the mezzanine floor. Belotserkovskiy demanded to know who Gogish was. Zhivotovskiy could not offer him any assistance. Zhivotovskiy asked source to spell the name Gogish. While writing the name down, Belotserkovskiy demanded to know if Gogish had personally told source that Maykapar was involved with the Rocketoplan. Source waited for Belotserkovskiy to cool down and then explained that it was Gogish's concept that was applicable to the Rocketoplan. However, since the short nozzle probably had a high divergence angle at the nozzle exit plane, source speculated that the Soviet Rocketoplan would experience flow separation as shown:

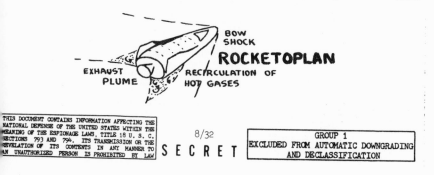

8/32

S E C R E T

Source stated that since Maykapar's paper at the 19th IAF Congress was concerned with the flow problem of a lifting reentry vehicle, it was natural to assume that Maykapar was involved in the problem. Belotserkovskiy abruptly terminated the discussion. As he was getting up from his seat, Vladimir Sychev approached the group huddled around source. Members of the group asked Sychev if he knew Gogish. Sychev nodded that he did and sat down in Belotserkovskiy's seat. Source reiterated the past discussion for Sychev's benefit. Source stated that his calculations indicated that the Gogish nozzle was approximately 50-70% shorter than a standard US bell nozzle. Sychev nodded his head in agreement and said, "You are right." Source then drew the indicated sketch below and indicated that as the nozzle is truncated toward the throat, nozzle performance (thrust

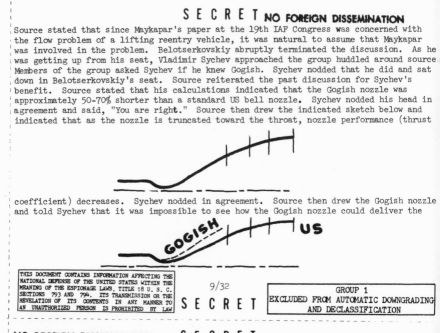

coefficient) decreases. Sychev nodded in agreement. Source then drew the Gogish nozzle and told Sychev that it was impossible to see how the Gogish nozzle could deliver the

9/32

S E C R E T

GROUP 1
EXCLUDED FROM AUTOMATIC DOWNGRADING
AND DECLASSIFICATION

NO FOREIGN DISSEMINATION S E C R E T

same performance as the longer US bell nozzles. Sychev replied that source's conclusion was correct and stated that Gogish's paper had not taken into consideration the specific thrust (specific impulse, ISP) losses. Source asked if the Gogish nozzle delivered approximately 2 or even 3% less performance than the longer US bell nozzle. Sychev stated that the assumption was correct. Source asked Sychev whether the Soviets were also considering rocket propulsion for both stages of their reusable launch vehicle. Sychev, who identified himself as the deputy director of the Central Aerohydrodynamics Institute (TsAGI), stated that rocket propulsion would be used in the upper stage (Rocketoplan) and airbreathing/rocket propulsion (composite engines) would most likely be used in the first stage. He stated that TsAGI was primarily responsible for the first stage and "the rocket people are working on the second stage." When source asked whether pure airbreathers (no rockets) would be used on the first stage, Sychev indicated that rockets would be needed in the first stage in some form or another.

SCRAMJET: Sychev asked source whether Professor Antonio Ferri (supersonic combustion specialist from New York University) was going to attend the conference. Source was not sure and used the question to mention that Professor Georgiy Petrov, director of the Institute of Space Research, had told source earlier this year that the Soviets were conducting research on a hydrogen SCRAMJET. Sychev stated that this was true. One system in particular that Sychev mentioned would collect air during the boost phase, liquefy it, and (1) use it in the first stage and (2) transfer some of the liquid air to the second stage for use later in the mission. Source notes that Petrov also mentioned a hydrogen SCRAMJET system that used liquid air and believes that the Soviet effort in this area is worth noting.

10/32

S E C R E T

GROUP 1
EXCLUDED FROM AUTOMATIC DOWNGRADING
AND DECLASSIFICATION

SPACE PROPELLANTS: Yuri Zavernaev identified himself as a specialist working with ablative materials for rocket nozzles at Moscow State University. Source asked Zavernaev whether he had worked with nitrogen tetroxide (N_2O_4) and he said that he had. Source asked whether nitrogen tetroxide - nitric acid oxidizers and kerosene fuels were attractive for near-Earth space applications. Zavernaev stated that they would be good in space.

INSTITUTE OF SPACE RESEARCH: Source had discussion with Gennadi Dimentiev, Igor Prissevok, and Jouli Khodarev. Igor Prissevok acted as translator on behalf of Dimentiev and Khodarev. During the discussion Prissevok referred to both men as Professor and comrade. Khodarev identified himself as deputy director of the Institute of Space Research, Moscow (G. Petrov is the Director.) Khodarev stated that his institute is responsible to the Academy of Sciences and was created to organize the institutes of the Academy. The institutes of the Academy are responsible to Petrov's institute as is shown below:

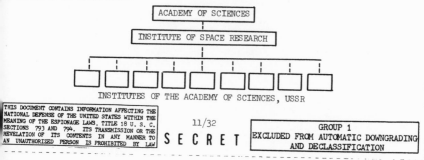

INSTITUTES OF THE ACADEMY OF SCIENCES, USSR

11/32

GROUP 1
EXCLUDED FROM AUTOMATIC DOWNGRADING AND DECLASSIFICATION

hodarev stated that his institute conducts research including rocket research.

SPACE LABORATORIES: When source asked if the Institute of Space Research would be ctive with Manned Orbital Laboratory-type programs, Khodarev stated that the primary bjective of the Institute was to plan experiments for orbital laboratories. Source sked whether a 50-100 man space station was too ambitious. Khodarev jokingly remarked hat it was small compared to the 2001 Space Odyssey movie. He added that the 2001 pace Odyssey film was "beautiful technically, but exaggerated too much."

TRI-PROPELLANT SYSTEMS: Professor Khodarev expressed considerable interest in the ithium-fluorine-hydrogen tripropellant system and asked whether the US was considering his system for a kick stage-type application. He mentioned that Soviet work indicated hat 510 seconds specific thrust (ISP) could be achieved. He mentioned the Grand Tour ission and asked whether we would modify the TITAN launch vehicle to accommodate a TTAN/CENTAUR/KICK STAGE configuration, using the tri-propellant system in the kick tage (kick stage is source's terminology) for this mission.

ADAPTABILITY OF ROCKETOPLAN TO PROTON BOOSTER: Source asked Khodarev whether the oviet Pocketoplan could be adapted to the Proton launch vehicle. Source added that ne Rocketoplan could be flight-tested in this manner if the Proton booster stages uld be adapted to the Rocketoplan. Khodarev understood the question, and gazed at the iling for approximately 20 seconds, as if he were conducting a mental exercise, and ated, "It can be done, but it would require considerable changes."

REUSABLE LAUNCH VEHICLES: Professor Khodarev was asked by source whether the viets would use rockets in both stages of their reusable space transport system (STS). odarev stated that the first stage would use airbreathing engines and the second stage uld use rocket engines. He added that this system "is better" than a rocket-rocket stem.

12/32

GROUP 1
EXCLUDED FROM AUTOMATIC DOWNGRADING AND DECLASSIFICATION

HIGH CHAMBER PRESSURE ROCKET ENGINES: ███████████████ remarked that he does
not like a rocket-rocket system. ███████████ was asked by source what rocket
engine chamber pressures would he use if he were building the US system. He responded,
"Very high chamber pressures are necessary." Source asked, "how about a 2000 psia bell
nozzle?" ███████████ stated that 2000 psia was no good.████████████████asked source
if US rocket specialists thought that 3000 psia was the highest rocket chamber pressure
in the world (Source notes that██████████was referring to the Pratt & Whitney 3000 psia
O_2-H_2 rocket engine). Source responded that there was no question that the US led the
world with a 3000 psia chamber pressure rocket engine.███████████said, "No...we have
higher." Source asked what the chamber pressure value was. ███████████shrugged his
shoulders indicating that he would not say. Source rephrased the question, "What is
the maximum chamber pressure rocket engine you would consider over the next 10 years?"
████████████still would not reply. Source rephrased question again, "What is the maximum
theoretical chamber pressure rocket engine you would consider over the next 10 years?"
████████████thought it over and said, "350 atmospheres." (approximately 5000 psia.)
Source quickly cut in, "Oh, you're talking about storable propellants...hydrogen is much
harder to pump." ████████████ cut in, "No, oxygen-hydrogen...oxygen-hydrogen." Source
added, "Transpiration cooling?"████████████ replied, "Yes...for high chamber pressure is
necessary."

LARGE LAUNCH VEHICLE: Source asked Professor Khodarev whether the Soviets had
considered a launch vehicle using a truncated plug with either an annular throat or
multiple chambers. Khodarev stated that they have. He mentioned that the multiple
chamber plug was easier to test. Source asked if they considered the single-stage-to-
orbit concept. Khodarev laughed and said, "If you have the money."

13/32

S E C R E T

S E C R E T

REUSABLE LAUNCH VEHICLES - PAYLOADS - MASS FRACTIONS: Igor Prissevok interpreted
for Gennadi Dimentiev and source. Dimentiev expressed considerable interest in the
Martin Marietta two-stage reusable space transport system. When source informed
Dimentiev that NASA was interested in payloads in the 50,000 lb. range, Dimentiev
seemed very concerned and asked whether the payload value was correct. Dimentiev gave
the impression that the Soviets would not consider a reusable space transport system
unless considerably more payload was involved. Source pursued the point and was able
to determine that Dimentiev was also concerned with the low propellant mass fraction,
lambda prime, (ratio of propellant weight/(propellant and inert weight) of the last
reusable stage. Source told Dimentiev that some recent US studies have indicated that
lambda primes of last stages are approximately 0.7. Dimentiev told source that if the
US could not design a reusable upper stage with a lambda prime of 0.8 or greater, then
the US should forget the whole concept. He was adamant in his position and understood
the definition of lambda prime. To improve the payload of the system, Dimentiev also
suggested that the US change from rocket engines to airbreathing engines in the first
stage. As the discussion progressed Dimentiev called V. Sychev for a consultation.
After explaining the problem to Sychev, Sychev reiterated that if the US were to use a
SCRAMJET/ROCKET first stage and a rocket second stage, the US would be able to deliver
a much higher payload into orbit. Source mentioned that in the past, Soviet cosmonauts
have referred to a lift reentry vehicle with movable or variable wing geometry. Source
asked Sychev if they were working on such a configuration. Sychev stated that such a
configuration is theoretically good, but practical engineering problems must be overcome
FLUORINATED OXIDIZERS, AIR POLLUTION: Source asked whether the Soviets were interested in fluorinated oxidizers for use in a reusable launch vehicle. Dimentiev commente█

14/32

S E C R E T

that fluorinated oxidizers were not favorable due to their toxicity. He also mentioned that during an abort it would be necessary to dump the propellant overboard.

HIGH CHAMBER PRESSURE OXYGEN-HYDROGEN STAGED COMBUSTION ENGINE CYCLE: Source showed ▨▨▨ a schematic of an engine cycle where a third fluid (not the fuel or oxidizer) operated in a closed system to drive the turbine. Source asked if the Soviets were using such a cycle and what the third working fluid was. ▨▨▨ stated that lithium was a good candidate for such a cycle. He added that such a cycle was old and the Soviets were using more advanced engine cycles now. Source asked if they had an oxygen-hydrogen preburner cycle (staged combustion) rocket engine. He answered yes. To make sure ▨▨▨ understood the question, source sketched the following engine cycle.

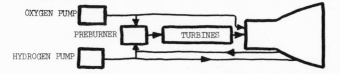

▨▨▨ nodded and pointed, saying, "Yes, yes," as if insulted that source had to draw cycle for him. Source asked again, "Oxygen-Hydrogen?" ▨▨▨ answered yes. Source asked, "How about nitrogen tetroxide at high chamber pressure?" ▨▨▨ emphatically stated, "No!" He indicated that N_2-O_4 was too dangerous to work with at high chamber pressures and sited its toxicity.

GROUP 1
EXCLUDED FROM AUTOMATIC DOWNGRADING
AND DECLASSIFICATION

ROCKETOPLAN - TANK DESIGN OPTION: The discussion between source, ▨▨▨, rissevok, and Dimentiev centered about the Rocketoplan, with Prissevok acting primarily as a translator. Source noted that when Dimentiev sketched a schematic of a lifting body in ▨▨▨ notebook, Dimentiev drew two rocket engines instead of one. Source believes that this could be significant, since source always drew one engine in his lifting body schematics up to that time.) Dimentiev asked whether the propellant tanks for the Rocketoplan would be integrated with the vehicle's wall-structure or whether special propellant tank containers would be designed in addition to the vehicle's wall-structure. His question appeared related to the lambda prime controversy mentioned earlier. During the discussion on propellant tank design, Dimentiev mentioned that one of the propellants could be stored in a tank surrounded by the other propellant. He offered this data as a method for reducing boiloff loses.

SPACE STORABILITY OF PROPELLANTS: Source asked Dimentiev: "For a space application in a near-Earth orbit, what is more important...specific thrust or storability?" Dimentiev did not hesitate to answer, "Storability." He implied that he was also concerned with reliability.

DESIGN APPROACH FOR A NOVA CLASS BOOSTER: Dimentiev asked source whether the US could build a NOVA class booster. Also, if we were to do so, would we develop new higher performing rocket propulsion systems sized to a new launch vehicle or would we use existing hardware, such as a cluster of Saturn V S-1C stages? Source noted that with Soviet rocket engines operating at consistently higher chamber pressures than comparable US engines, the incentive for the Soviets to develop even higher performing rocket propulsion systems for larger launch vehicles is less than that for the US.

GROUP 1
EXCLUDED FROM AUTOMATIC DOWNGRADING
AND DECLASSIFICATION

ROCKET ENGINE ADVANCED MATERIALS, DESIGN: When asked what materials the Soviets use in their rocket engines to withstand the high heat fluxes, Dimentiev mentioned one nozzle system that used silicon impregnated in the metal. (Note: possibly SiO_2) About a minute later he mentioned a copper-chrome alloy. Source notes that cooper-chrome is suspected of being used for the nozzle of the RD-107 engine. Dimentiev added that the silicon additive is an old idea. With regard to advanced engine concepts, Dimentiev stated that the Soviets have looked at molybdenum and/or its alloys and found that molybdenum was too rigid after reuse. He suggested niobium was a better alternative. Dimentiev also indicated that tantalum was being considered for advanced nozzle concepts, making it clear that tantalum would be used near the nozzle exit plane rather than near the throat. Dimentiev stated that advanced Soviet engines that operate at high chamber pressures use relatively low contraction ratio combustion chambers (relative to the high contraction ratio Soviet engines designed during the 1950's.) When asked whether the Soviets used additives in their propellants, Dimentiev stated that this was a question for his students who were concerned with theoretical investigations, not experimental. Source then asked if the Soviets were using small percentages of ozone in their oxidizers He responded with, "No," but agreed with source that small concentrations of ozone in an oxidizer increases the combustion rate and permits the use of smaller combustion chambers

LARGE LAUNCH VEHICLE DESIGN: Source drew a sketch of a plug, as shown below, and discussed the multichamber or annular throat plug engine for advanced large launch vehicles. During the discussion Dimentiev appeared slightly disturbed at source's sketch and made two changes. First, he rounded off the base of the plug as shown below and then sketched the base of the launch vehicle's propellant tank so that it partially occupied the inner volume of the plug.

17/32

NO FOREIGN DISSEMINATION

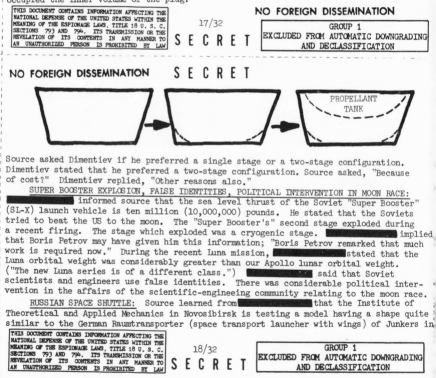

PROPELLANT
TANK

Source asked Dimentiev if he preferred a single stage or a two-stage configuration. Dimentiev stated that he preferred a two-stage configuration. Source asked, "Because of cost?" Dimentiev replied, "Other reasons also."

SUPER BOOSTER EXPLOSION, FALSE IDENTITIES, POLITICAL INTERVENTION IN MOON RACE: ▮▮▮▮▮▮▮▮▮▮ informed source that the sea level thrust of the Soviet "Super Booster" (SL-X) launch vehicle is ten million (10,000,000) pounds. He stated that the Soviets tried to beat the US to the moon. The "Super Booster's" second stage exploded during a recent firing. The stage which exploded was a cryogenic stage. ▮▮▮▮▮▮▮▮▮▮ implied that Boris Petrov may have given him this information; "Boris Petrov remarked that much work is required now." During the recent Luna mission, ▮▮▮▮▮▮▮▮▮▮ stated that the Luna orbital weight was considerably greater than our Apollo lunar orbital weight. ("The new Luna series is of a different class.") ▮▮▮▮▮▮▮▮▮ said that Soviet scientists and engineers use false identities. There was considerable political intervention in the affairs of the scientific-engineering community relating to the moon race.

RUSSIAN SPACE SHUTTLE: Source learned from▮▮▮▮▮▮▮▮that the Institute of Theoretical and Applied Mechanics in Novosibirsk is testing a model having a shape quite similar to the German Raumtransporter (space transport launcher with wings) of Junkers in

18/32

their wind tunnel.

MORE SCIENCE CITIES; TREATMENT OF FOREIGNERS: ▓▓▓▓▓▓▓▓ said that there will be five (5) more Science Cities built in Siberia. One is almost completed and is located in or near a wooded area. He indicated that the Soviets are having difficulties in getting people (young students included) to go to Novosibirsk. The students claim that there is nothing to do, no entertainment. One student was reported to have said, "Do you want me to work in a place of nothing!" ▓▓▓▓▓▓▓▓ said that foreigners get the best seats and first choice in Soviet commercial aircraft because the Soviets are interested in the dollar. He added that people who live in the USSR are not free to take an airplane trip; travel is forbidden.

PERSONNEL: ▓▓▓▓▓▓▓▓ reiterated that he worked at ▓▓▓▓▓▓▓▓ (▓▓▓). Source asked whether ▓▓▓ knew Professor Borisenko, whom source had met in 1966. ▓▓▓▓▓ stated that Borisenko was still ▓▓▓▓▓▓ for G. G. Cherniy. Source asked about Professor G. L. Grodzovskiy, who was unable to attend the conference. It was learned that Grodzovskiy is now working at TsAGI with V. Sychev. Source asked if Grodzovskiy was Sychev's associate. ▓▓▓▓▓▓▓ replied that the term "associate" in the USSR meant someone who worked for another, or a subordinate, and one could not classify Grodzovskiy in that category. When source asked whether Grodzovskiy was on the same level as Sychev, ▓▓▓▓▓▓ did not answer and did not want to discuss the matter anymore. He did mention that Grodzovskiy also teaches at MAI. Source believes that Grodzovskiy is now spending more time at TsAGI than the Moscow Physical Technical Institute. Source met Youri V. Zonov, who could easily pass as an American engineer from New York City. Zonov stated that he was raised in England, where his father was ambassador/worked at the embassy. He was assigned to the UN, where he helped draft the space treaty. He

19/32

S E C R E T

GROUP 1
EXCLUDED FROM AUTOMATIC DOWNGRADING AND DECLASSIFICATION

NO FOREIGN DISSEMINATION S E C R E T

stated that his wife was preparing a paper on the US military-industrial complex and how it affects our economy, political objectives, etc. Her paper concludes that the complex exists and the system (military-industrial complex - US economy - US politics, etc) perpetuates itself. When asked whether he had a solution to the US racial problem, he stated that the USSR has their own problems internally. Source asked him to elaborate but he would not. Source believes that his statement: "We have our own problems," is significant because he received the identical response over the same question from Professor Zhivotovskiy earlier in the year. When asked about the space race, Zonov remarked that the Komarov accident slowed down their space program considerably and cost them the race. He noted that the Soviets use a step wise approach in their space program and many projects were held up because of the accident. Zonov currently works at the Institute of Space Research and mentioned that he is chief of the Scientific Information Department. Zonov indicated that he does not have a good scientific or technical background,"but is strong on languages."

INSTITUTE OF SPACE RESEARCH: Source and ▓▓▓▓▓▓▓▓ discussed the Institute of Space Research with Deputy Director Khodarev. Khodarev stated that the members of the USSR Academy of Sciences selected Professor Georgiy Petrov as Director of the newly created Institute of Space Research. The members of the Presidium of the USSR Academy of Sciences selected the deputy director (himself). The Presidium also selected the department heads of the Institute. The department heads were free to select their own personnel, with Georgiy Petrov having veto power. L. Bashckirov stated that he works at the Institute of Space Research and is concerned with the transmission of data between spacecraft and Earth.

TRI-PROPELLANTS: In the presence of ▓▓▓▓▓▓▓ Professor Khodarev again raised

20/32

S E C R E T

GROUP 1
EXCLUDED FROM AUTOMATIC DOWNGRADING AND DECLASSIFICATION

the question of a Titan/Centaur/Kick Stage, using Lithium-Fluorine-Hydrogen in the kick stage. Source notes that this was the third time Professor Khodarev brought up the subject of tripropellants. Using Youri Zonov as his interpreter, Khodarev mentioned the 1974-1978 time period as possible mission applications.

DIMENTIEV APPARENTLY AN IMPORTANT MAN: ████████████████████████ stated that Gennadi Dimentiev was a "Big Shot" and that source didn't know how big Dimentiev was. He also called him comrade and told source that Dimentiev had privately given Dr. George Mueller of NASA a personal gift from the Soviet cosmonauts to be delivered by Mueller to the American astronauts.████████████ added that Dimentiev "works with rockets and cosmonauts."

SOVIET COSMONAUTS - LUNAR LANDING: Source asked Prissevok how many cosmonauts were in training. Prissevok thought the question over and said "at least 25 or 30 cosmonauts." Source asked if he knew the names of the cosmonauts who would land on the moon. Prissevok stated that the names have not been selected yet, but a decision was forthcoming. When source asked whether he preferred the direct lunar descent mode or a lunar orbit and then lunar descent, Prissevok stated that he preferred a circular lunar orbit before descent. Source asked about Aleksei Leonov. Prissevok stated that Leonov has not retired and is still training. With regard to Professor Khodarev, Prissevok stated that the Professor escorted US Astronaut Frank Borman during Borman's Soviet tour. Later in the evening Prissevok remarked that Professor Khodarev was "trying to buy" him so that he could work under the Professor at the Institute of Space Research. Prissevok stated that he was in Athens during the 1965 conference and personally escorted Cosmonauts Leonov and Belyayev.

21/32

ZHIVOTOVSKIY-BIOGRAPHICAL: Source Learned the Following Information Concerning Professor Zhivotovskiy From████████████: Professor Zhivotovskiy pays 16 rubles for his flat in Moscow, which includes 3 rooms. The Professor said that when he was young he used to dance. However, when he got married his wife, who does not care to dance, ██. He remarked that he practices modern dances in private (behind a closed door) for exercise. The Professor asked about the "special" airliner that brought ████████ to Argentina. "Number one" showed the Professor a photographic album that contained pictures of "number one's" last trip. The Professor shuffled through the pictures until he got to pictures of "number one's" spouse's relatives in Greece. The Professor stopped, leafed back and looked at the relatives again. On his return to Russia, the Professor indicated that he may stop in Dakar and go big-game hunting for two days. He indicated that he would stay in Mar del Plata until Sunday 12 October 1969.

INSTITUTE OF SPACE RESEARCH: Professor Khodarev stated that the Institute of Space Research building is not yet completed. The building will be approximately $\frac{1}{2}$ km. long and eleven stories high. The institute is located approximately three km. from Moscow State University (MGU). Professor Khodarev has the benefit of the institute automobile, driving ten minutes to work each day. He noted that Professor Georgiy Petrov lives between the Institute of Space Research and Moscow State University, where he also teaches. Igor Prissevok stated that Georgiy Petrov "always has time for students." Petrov's favorite hobby is fishing.

COMMUNICATIONS - TELEVISION - FILM: Professor Khodarev stated that he was responsible for the television communication system of Luna, Zond 3, Luna Orbiters, and Luna soft landers. Source asked about Zond 5 and 6. Khodarev stated that both systems

22/32

used film that was recovered. He stated, "Zond 5 and Zond 6 each took 100 and...about 100 pictures." Khodarev stated that large negatives were used to obtain high resoultion (he indicated that the size of each negative was approximately 9" x 9", using his hands as a frame.) ████████████████████████ stated that Soviets had invented a film that could withstand 5×10^8 neutrons/cm^2 without being spoiled. He remarked that they also have invented a chemical solution that permits film to be exposed to ordinary light prior to normal camera usage. The chemical solution is applied after normal exposure and prior to developing.

SERGEI P. KOROLEV: Source asked Professor Khodarev if he knew the late Soviet rocket designer, S. P. Korolev. Professor Khodarev stated that he worked with Sergei Korolev for well over ten years and knew him on a close personal basis. At this point source noted that Khodarev became emotional and detected tears in his eyes. He continued, saying that he saw Korolev three weeks before he died, in January. Korolev was in a hospital for six days. "They operated on him...and, he died,...stomach cancer." Khodarev stated that Korolev's death was a great shock to the entire community, as he appeared healthy prior to being admitted to the hospital. Khodarev described Korolev as "number one." Source mentioned that Korolev had designed the Vostok launch vehicle. Khodarev added that Korolev was also responsible for and designed Voshkod and Zond-3, and was "coordinating the design of the Proton launch vehicle." Khodarev also mentioned that Georgiy Petrov was a personal friend of Korolev and worked with him as well.

LUNAR BASES: Source asked Khodarev about lunar bases. Khodarev replied, "We have plans." He added that the major problem under current consideration was concerned with the large thermal gradients that one encounters on the lunar surface. He mentioned

GROUP 1
EXCLUDED FROM AUTOMATIC DOWNGRADING AND DECLASSIFICATION

telescope mirrors and construction material of the station (lunar base.)

PLASMA ROCKETS - NUCLEAR PROGRAM: ████████████████ implied that plasma rocket research was receiving more emphasis than nuclear rocket propulsion. In the discussion he mentioned a nuclear-electric rocket. He also mentioned the Soviet experience with the YANTAR plasma rocket for ACS. ████████ agreed with the source that plasma rockets could be used for an Earth-Moon space shuttle; interplanetary transportation; near-Earth orbital delta velocity and plane change delta velocity requirements.

PROPELLANT OXIDIZERS: In a discussion on propellant oxidizers, source asked ████████ about nitrogen tetroxide (N_2O_4). ████████ stated, "mixed with fluorine is best". Source asked if he was referring to FLOX or OF_2, or combinations of oxygen and fluorine. ████████ said no. Source and "number two" wrote $N_2O_4 + F_2$ on a napkin. ████████ gave no reply. Source and "number two" wrote N_2F_4 on the napkin. ████████ smiled, but didn't say anything. Source and "number two" said that N_2F_2 was the answer. ████████ sportingly laughed with source and "number two" as if to acknowledge they were right, but would not say it directly. ████████████████

KHODAREV - PETROV: Professor Khodarev said that he was born near the Baku region. He has a sister in Moscow and a retired uncle in Novosibirsk. Igor Prissevok mentioned that Professor Georgiy Petrov has a son and daughter, both over six feet tall. With regard to his travel schedule, Khodarev has already visited Leningrad nine times this year, using the morning train on a six hour run. He stated that he plans to visit the US in 1970 during January to attend the AIAA meeting and/or in June-July to visit Cape Kennedy. He may contact source during this visit.

GROUP 1
EXCLUDED FROM AUTOMATIC DOWNGRADING AND DECLASSIFICATION

ZHIVOTOVSKIY - PRISSEVOK: Source noted that Igor Prissevok used the word "chap" to describe an individual. Source recalls that Professor Zhivotovskiy used the word "chap" freely, on at least four separate occasions, and speculates that Prissevok may have considerably more contact with Zhivotovskiy than he lets on.

LUNAR PROGRAM - WEIGHTS IN ORBIT REQUIRED: ▓▓▓▓▓▓▓▓▓▓▓ stated that Luna went into an orbit about the moon to check the landing system/sequence, using orbital plane changes.▓▓▓▓▓▓▓▓stated that the Soviets now believe that an orbit(s) around the moon is necessary prior to descent; a direct landing is too dangerous. He continued, a two (2) stage system will be used; one for descent, one for ascent. He pointed out that the direct mode required more propellant. The Soviet system that will be used requires 150 tons in an Earth orbit and 20-30 tons in moon orbit. Shortly after his exchange with▓▓▓▓▓, source asked ▓▓▓▓▓▓▓ what the conversion was between miles and kilometers. ▓▓▓▓▓▓cut in and gave the answer. Then source asked the conversion between liters and quarts. ▓▓▓▓▓▓▓▓▓ responded. Source finally asked "tons to pounds."▓▓▓▓▓responded 2.2 x 10^3 pounds per ton. Source feels certain that▓▓▓▓▓ figures represented metric tons (150 tons or 330,000 lb; 20-30 tons or 44,000 - 66,000 lb.)

SL-X LAUNCH VEHICLE: ▓▓▓▓▓▓▓▓▓▓▓▓▓▓▓▓acknowledged that the Soviets have a launch vehicle that is larger than the Saturn V,▓▓▓▓▓▓▓reluctant to discuss any details.

INTERCOSMOS: ▓▓▓▓▓reported that the Intercosmos organization, which is headed by Soviet Academician Boris Petrov, held a secret meeting in East Berlin during May 1969. He stated that the primary purpose of the meeting was to solve managerial and organizational problems. Source asked whether Intercosmos dictated projects to the

THIS DOCUMENT CONTAINS INFORMATION AFFECTING THE NATIONAL DEFENSE OF THE UNITED STATES WITHIN THE MEANING OF THE ESPIONAGE LAWS, TITLE 18 U. S. C. SECTIONS 793 AND 794. ITS TRANSMISSION OR THE REVELATION OF ITS CONTENTS IN ANY MANNER TO AN UNAUTHORIZED PERSON IS PROHIBITED BY LAW

S E C R E T

GROUP 1
EXCLUDED FROM AUTOMATIC DOWNGRADING AND DECLASSIFICATION

satellite country institutes. ▓▓▓▓▓stated that the satellite institutes were free to propose new ideas to Boris Petrov and that the Soviets did not dictate projects in an absolute manner.

CHINA - CHOU EN-LAI - LIN PIAO: Horst Hoffmann stated that they hoped that Mao Tse-tung would pass away (Hollax jokingly pointed to Heaven), as things would change for the better with Chou En-lai. Source asked about Lin Piao, but Hoffmann said that Chou En-lai had the real power, but was afraid to assert himself with Mao still living. Hoffmann added that he personally knew/met Chou En-lai fifteen years ago in Berlin. He reiterated that Chou En-lai had a slightly different philosophy than Mao, but could not promulgate his ideas as long as Mao was in power.

IGOR PRISSEVOK - KGB WATCHDOG: Prissevok told source that he was assigned the responsibility of picking up everyone's passports so that they could "take care of the necessary paperwork" before they departed. Source asked how many passports were involved. Prissevok responded "24...25...28." Source used Prissevok's current task to inquire about Igor Milovidov (former KGB watchdog at these conferences who was usually registered as the scientific secretary.) Prissevok stated that Milovidov was replaced Yuri Barinov. Source asked if Milovidov was promoted. Prissevok did not respond. Source asked if Barinov was attending the conference. Prissevok stated that Barinov was replaced by another man, whose name he could not recall. Source feels that Prissevok holds a special position because (1) he acted as an interpreter most of the time, (2) he tried to change the subject when source and Professor Khodarev discussed Soviet programs and personnel, (3) he escorted the Soviet cosmonauts in Greece, (4) he was in charge of Soviet passports, (5) he talked like Zhivotovskiy, using the term "chap", (6) he received orders from Zhivotovskiy on one occasion, and (7) he thought

THIS DOCUMENT CONTAINS INFORMATION AFFECTING THE NATIONAL DEFENSE OF THE UNITED STATES WITHIN THE MEANING OF THE ESPIONAGE LAWS, TITLE 18 U. S. C. SECTIONS 793 AND 794. ITS TRANSMISSION OR THE REVELATION OF ITS CONTENTS IN ANY MANNER TO AN UNAUTHORIZED PERSON IS PROHIBITED BY LAWNO FOREIGN DISSEMINATION

S E C R E T

GROUP 1
EXCLUDED FROM AUTOMATIC DOWNGRADING AND DECLASSIFICATION

nothing of accompanying source on a two hour walk around Mar del Plata. Source feels that most Soviet delegates would hesitate to walk through a city with foreigners unless they were accompanied by other members of their delegation or a watchdog. Source feels that Prissevok therefore holds a special position and was reporting to Zhivotovskiy after each day. Source feels that it was Prissevok who informed Zhivotovskiy between 1:00 a.m. - 8:00 a.m. on Wednesday about the previous night's drinking party. Prissevok has one daughter who is 7½ years old. His wife is an economist, and works. Igor's in-laws take care of the daughter, who does not care for school. His wife may quit her job because of this current problem. It takes Igor approximately 45 minutes to get to work. He lived close to the Chinese border in Siberia when he was a child. Igor stated that he has been in Denver and California. Igor remarked that it was very difficult for the Soviets to get into Argentina, since the Argentine military regime is very anti-Communist.

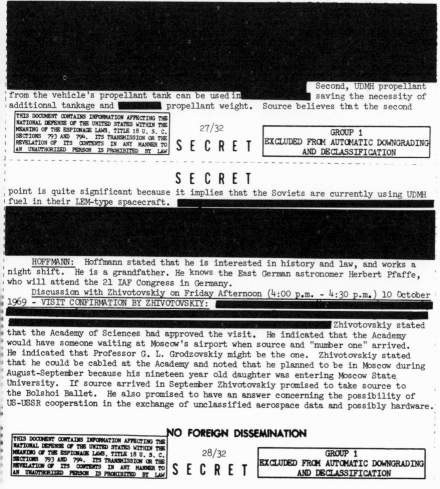

from the vehicle's propellant tank can be used in █████████ Second, UDMH propellant █████████ saving the necessity of additional tankage and ████████ propellant weight. Source believes that the second

27/32

S E C R E T

S E C R E T

point is quite significant because it implies that the Soviets are currently using UDMH fuel in their LEM-type spacecraft. █████████████████

HOFFMANN: Hoffmann stated that he is interested in history and law, and works a night shift. He is a grandfather. He knows the East German astronomer Herbert Pfaffe, who will attend the 21 IAF Congress in Germany.

Discussion with Zhivotovskiy on Friday Afternoon (4:00 p.m. - 4:30 p.m.) 10 October 1969 - VISIT CONFIRMATION BY ZHIVOTOVSKIY: █████████████████

███████████████████████████ Zhivotovskiy stated that the Academy of Sciences had approved the visit. He indicated that the Academy would have someone waiting at Moscow's airport when source and "number one" arrived. He indicated that Professor G. L. Grodzovskiy might be the one. Zhivotovskiy stated that he could be cabled at the Academy and noted that he planned to be in Moscow during August-September because his nineteen year old daughter was entering Moscow State University. If source arrived in September Zhivotovskiy promised to take source to the Bolshoi Ballet. He also promised to have an answer concerning the possibility of US-USSR cooperation in the exchange of unclassified aerospace data and possibly hardware.

28/32

S E C R E T

ZHIVOTOVSKIY PERSUASION: Professor Zhivotovskiy contacted source, Professor M. Barrere of France, Professor Andre Jaumotte of Belgium, and ▮▮▮▮▮▮▮▮▮ to make sure that all helped to persuade Dr. Partel to hold the 3rd International Conference on Space Engineering in Florence, Italy during 1971.

Observations Made Friday Evening (11:35 p.m.), 10 October 1969 and Saturday Afternoon (12:00 - 1:00 p.m.) 11 October 1969: Source observed Professor Zhivotovskiy and Professor Oleg Gazenko holding a private discussion on the 4th floor of the Hotel Provincial near the elevator at 11:35 p.m. Source thought nothing of it until he noticed them again at 12:00 a.m., almost a half hour later. They appeared concerned over something and it appeared to source that Zhivotovskiy was giving Gazenko instructions. Source feels that it would be very worthwhile if it can be established who sat next to Gazenko during the annual banquet, which was held between 9:00 p.m. and 11:15 p.m. prior to the Zhivotovskiy-Gazenko meeting on 10 October 1969. On the following day source observed Zhivotovskiy leave the Hotel Provincial at 12:25 p.m. and walk in the general direction of the Soviet intelligence headquarters. At 12:30 p.m. source observed Gazenko in the general vicinity of their headquarters walking in the opposite direction (See map below.)

29/32

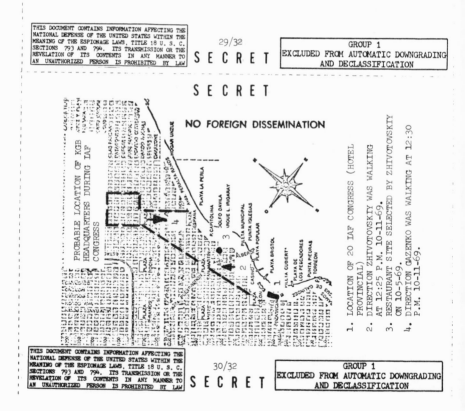

1. LOCATION OF 20 IAF CONGRESS (HOTEL PROVINCIAL).
2. DIRECTION ZHIVOTOVSKIY WAS WALKING AT 12:25 P.M. 10-11-69.
3. RESTAURANT SITE SELECTED BY ZHIVOTOVSKIY ON 10-5-69.
4. DIRECTION GAZENKO WAS WALKING AT 12:30 P.M. 10-11-69.

PROBABLE LOCATION OF KGB HEADQUARTERS DURING IAF CONGRESS

30/32

ISOLATED NOTES: Source believes that the Soviets are "aware" of ▆▆▆▆▆▆▆▆ and recommends that ▆▆▆▆▆▆▆▆▆▆▆▆▆▆▆▆▆▆▆. Igor Prissevok told source that the Soviets attend conferences because it is their best means of acquiring information. He indicated that receptions, cocktail parties, and similar gatherings were valued more than papers and presentations made during formal sessions at these conferences. Professor Zhivotovskiy told source that he thought that the Venice conference (2nd International Engineering Conference on Space Engineering - Venice, Italy - May 1969) was very worthwhile. He indicated that he was in Venice making preparations fifteen days before the conference started. Zhivotovskiy said he will provide a "strong" delegation in support of the 3rd International Engineering Conference on Space Engineering - Italy - Spring of 1971. Source suggests that Zhivotovskiy not be allowed to control this conference as he did the Venice conference. Source has verbal comments on this subject. Zhivotovskiy asked source to provide him with unclassified material relating to the storability of cryogenic propellants, particularly hydrogen, on Earth. He requested data on insulation materials, techniques, and methods of transporting cryogenics. He told source to use his own discretion in selecting the material.

COMMENTS RELATING TO INVITATION BY ZHIVOTOVSKIY: Source believes that he will meet G. L. Grodzovskiy, Y. Joulev, G. Maykapar, V. Sychev, G. Dimentiev, Y. Zavernaev, Y. Khodarev, and I. Prissevok in the Moscow area in addition to others he does not know personally. Zhivotovskiy indicated that source might be able to visit Yuri Moskalenko when in Leningrad. Source believes that he might get a chance to visit the following institutes when in the Moscow area. 1. Moscow State University; 2. Moscow Physical Technical Institute; 3. Institute of Space Research; 4. Academy of Sciences; 5.Central Aero-Hydrodynamics Institute; and others. Source asked if he could present a paper.

THIS DOCUMENT CONTAINS INFORMATION AFFECTING THE NATIONAL DEFENSE OF THE UNITED STATES WITHIN THE MEANING OF THE ESPIONAGE LAWS, TITLE 18 U. S. C. SECTIONS 793 AND 794. ITS TRANSMISSION OR THE REVELATION OF ITS CONTENTS IN ANY MANNER TO AN UNAUTHORIZED PERSON IS PROHIBITED BY LAW	31/32 S E C R E T	GROUP 1 EXCLUDED FROM AUTOMATIC DOWNGRADING AND DECLASSIFICATION

S E C R E T

Zhivotovskiy told source that he should feel free to present a paper(s) on subject(s) of source's own choosing. He promised source interpreters during all technical sessions. Source was promised an opportunity to visit the scientific institutes in Science City, near Novosibirsk and meet the personnel at these institutes. Earlier in the year Zhivotovskiy mentioned that source could visit the Institute of Nuclear Physics. At this conference, Zhivotovskiy gave the impression that the work was unclassified at most of the institutes in Science City. There is a chance that source may visit the Crimean Observatory on the return trip to Moscow. Zhivotovskiy promised source that he would have ample time to see the cultural attractions in the Soviet Union, in addition to the "business" portion of the visit.

THIS DOCUMENT CONTAINS INFORMATION AFFECTING THE NATIONAL DEFENSE OF THE UNITED STATES WITHIN THE MEANING OF THE ESPIONAGE LAWS, TITLE 18 U. S. C. SECTIONS 793 AND 794. ITS TRANSMISSION OR THE REVELATION OF ITS CONTENTS IN ANY MANNER TO AN UNAUTHORIZED PERSON IS PROHIBITED BY LAW	32/32 S E C R E T	GROUP 1 EXCLUDED FROM AUTOMATIC DOWNGRADING AND DECLASSIFICATION

Background: On 5 October 1969, Professor Georgiy Zhivotovskiy invited source to visit the Soviet Union in 1970 as guest of the Academy of Sciences. On 9 October 1969, Zhivotovskiy asked source for information pertaining to source's scheduled departure from Mar del Plata (flight number, date, and time).

Summary: On 11 October 1969 as source deplaned from his Mar del Plata-Buenos Aires flight, one of three Russians standing at the foot of the flight staircase, took source's picture. These three are referred to as the "Penguin Trio" and were met by three other Russians inside the terminal, referred to as "Embassy Types." Source took photographs of the Russians and noted their actions indicated that they were guarding very valuable luggage carried by the "Penguin Trio."

THREE USSR "EMBASSY TYPES"

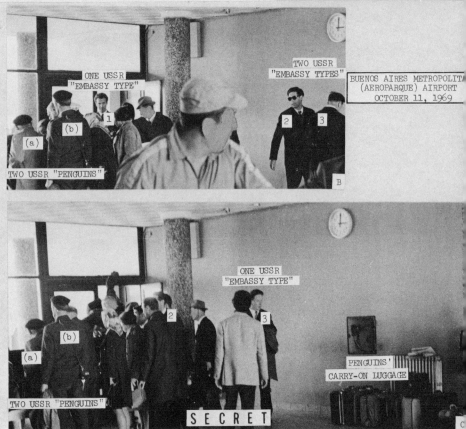

ONE USSR "EMBASSY TYPE"

TWO USSR "EMBASSY TYPES"

BUENOS AIRES METROPOLIT (AEROPARQUE) AIRPORT OCTOBER 11, 1969

TWO USSR "PENGUINS"

ONE USSR "EMBASSY TYPE"

TWO USSR "PENGUINS"

PENGUINS' CARRY-ON LUGGAGE

Appendix III

The officials, scientists, and engineers listed below are involved in the Soviet Space Shuttle Program:

1. *Vladimir A. Kirillin*, chairman of the State Committee for Science and Technology, is administratively involved in the development of the reusable booster stage and the Rocketoplan. He will ensure that the State Committee's facilities, institutes, and design bureaus are technically responsive to the directives of the Space Shuttle Commission and Kosygin's Council of Ministers.

2. *Mstislav V. Keldysh*, president of the U.S.S.R. Academy of Sciences, is administratively involved in the development of the reusable booster stage and the Rocketoplan. Keldysh will ensure that the academy's facilities, institutes, and design bureaus are scientifically and technically responsive to the directives of the Space Shuttle Commission and the Council of Ministers. When the Russian shuttle is completed, it will be managed by the Academy of Sciences for the Ministry of Defense.

3. *Georgiy I. Petrov*, director of the Institute of Space Research of the U.S.S.R. Academy of Sciences, is primarily the consultant for the development of both the reusable booster and the Rocketoplan. He is also involved in the development of the scramjet engines that are being designed for use in the reusable booster. Some of this work is being

done at the Institute of Mechanics at Moscow State University, but Petrov's responsibilities at the Institutite of Space Research will occupy most of his time.

4. *Boris N. Petrov*, director of the Institute of Automatics and Telemechanics, of the U.S.S.R. Academy of Sciences, is the consultant on space shuttle organizational and administrative matters. Petrov is technically involved in the development of the Rocketoplan's control system and is active at both the Moscow Aviation Institute and the Institute of Automatics and Telemechanics.

5. *G. L. Grodzovskiy*, one of the secret heads of the Russian space program, is technically involved in the development of both the reusable booster and the Rocketoplan, and is active at the Moscow Aviation Institute, Central Aerohydrodynamics Institute, Moscow Physical Technical Institute, and Moscow State University. He is also responsible for coordinating the integration of the rocket and jet propulsion systems to these vehicles.

6. *Gennadi Dimentiev* (pseudonym), one of the secret heads of the Russian space program, is technically involved in the development of the Rocketoplan and will be active with systems design problems at the Moscow Aviation Institute.

7. *Valentin P. Glushko*, chief designer of rocket engines, U.S.S.R. is responsible for the development of the high chamber pressure oxygen-hydrogen rocket engines that will power the Rocketoplan, and will be active at his design bureau.

8. *V. V. Struminsky*, former head of the Institute of Theoretical and Applied Mechanics near Novosibirsk, is administratively and technically involved in the development of the reusable booster and the Rocketoplan. His former institute will support the Central Aerohydrodynamics Institute with wind tunnel tests. He is currently affiliated with the Institute of the Problems of Mechanics, where he is coordinating activities relating to shuttle structural and materials problems.

9. *Oleg M. Belotserkovskiy*, director of the Moscow Physical Technical Institute, is administratively involved in the space shuttle effort and is consulted on major managerial decisions concerning the Rocketoplan and the assignment of personnel from his institute. He will also be active at the Computer Center of the U.S.S.R. Academy of Sciences.

10. *Vladimir Sychev*, deputy director of the Central Aerohydrodynamics Institute, is administratively and technically involved in the development of the reusable booster and is responsible for major

managerial and technical decisions concerning the design of the booster stage. He is also active at the Moscow Physical Technical Institute and is the chief liaison officer between the Central Aerohydrodynamics Institute and the U.S.S.R. Academy of Sciences on space shuttle matters.

11. *G. I. Maykapar*, a renowned aeronautical engineer in the Soviet Union, is technically involved in the development of both the reusable booster stage and the Rocketoplan, and is responsible for the design and wind tunnel testing of space shuttle scale models. He is very much involved in propulsion system and vehicle integration problems and is active at the Central Aerohydrodynamics Institute, Moscow Aviation Institute, and Moscow Physical Technical Institute.

12. *G. G. Chernyi*, director of the Institute of Mechanics of Moscow State University, is administratively and technically involved in the development of the reusable booster and the Rocketoplan. He is coordinating the wind tunnel experiments and research on scramjet engines in support of the reusable booster program, and is very much involved in the design of special rocket engine components, such as "short nozzles." This nozzle work complements V. P. Glushko's efforts to develop the rocket engines for the Rocketoplan.

The facilities listed below are actively involved in the development of the Russian space shuttle:

1. The Moscow Aviation Institute (MAI)
2. Central Aerohydrodynamics Institute (TsAGI)
3. Glushko Rocket Engine Design Bureau
4. Mikoyan Aircraft Design Bureau
5. Moscow Physical Technical Institute (MPTI)
6. Institute of Mechanics of Moscow State University
7. Computer Center of the U.S.S.R. Academy of Sciences
8. Institute of Automatics and Telemechanics of the U.S.S.R. Academy of Sciences
9. Institute of Space Research of the U.S.S.R. Academy of Sciences
10. Institute of Theoretical and Applied Mechanics, Novosibirsk

The men and facilities listed above contributed greatly to Soviet advanced development programs during the 1950s and 1960s, and are in no small way responsible for the Soviet's current powerful and competitive position in the aerospace field. Their work on the Russian Space Shuttle Program should be followed closely.

Bibliography

Abel, Elie, *The Missile Crisis* (Philadelphia: J. B. Lippincott Company, 1966).

Accoce, Pierre and Pierre Quet, *A Man Called Lucy* (New York: Coward McCann, Inc., 1966).

Barron, John, "The Schooling of a Soviet Spy," from *KGB* (New York: The Reader's Digest, April 1970).

Bernikow, Louise, *Abel* (New York: Trident Press, 1970).

Bourke, Sean, *The Springing of George Blake* (New York: The Viking Press, 1970).

Buckley, William F., Jr., *Up From Liberalism* (New York: William Morris Agency, 1959).

Carpozi, George, Jr., *Red Spies in Washington* (New York: Trident Press, 1968).

Clifford, Clark M., *A Statement by Secretary of Defense Clark M. Clifford: The 1970 Defense Budget and Defense Program for Fiscal Years 1970–1974.* (Washington: U.S. Government Printing Office).

Cohn, Roy, *McCarthy* (New York: The New American Library, 1968).

Crankshaw, Edward, *Khrushchev, A Career* (New York: The Viking Press, Inc., 1966).

Cresswell, Mary Ann and Carl Berger, *United States Air Force History* (Washington: Office of Air Force History, 1971).

Daniloff, Nicholas, *The Kremlin & The Cosmos* (New York: Alfred A. Knopf, 1972).

Devoe, Barbara, *Astronaut Information: American and Soviet* (Washington: The Library of Congress Legislative Reference Service, 1968).

Dulles, Allen, *The Craft of Intelligence* (New York: Harper & Row Publishers, 1963).

Fischer, Louis, *Fifty Years of Soviet Communism* (New York: Popular Library, 1968).

Gould, Kenneth M. and Richard E. Gross, *The Soviet Union* (New York: Scholastic Book Services, 1963).

Goldston, Robert, *The Soviets* (New York: Bantam Books, 1967).

Hoover, J. Edgar, *Masters of Deceit* (New York: Henry Holt and Company, 1958).

Hoover, J. Edgar, *On Communism* (New York: Random House, 1969).

Huminik, John, *Double Agent* (New York: The New American Library, 1967).

Huss, Pierre J. and George Carpozi, Jr., *Red Spies in the U.N.* (New York: Coward-McCann, Inc., 1965).

Jackson, Senator Henry M., "Strategic Arms Limitation Talks (SALT): Legislative History of the Jackson Amendment, 1972" (news release from the office of the Senator), full record of Congressional debate on the ABM Treaty and the resolution authorizing approval of the Interim Agreement on Offensive Weapons (*Congressional Record*, August 3, 1972 through September 25, 1972).

Jackson, Senator Henry M., "The Senate and the Interim Agreement: Statement by Senator Henry M. Jackson" (news release from the office of the Senator), delivered on the Senate floor on August 11, 1972.

James, Peter N., *Annual Report on Foreign Rocket Technology (1968–1969)*, PWA-FR-3195 (West Palm Beach, Florida: Pratt & Whitney Aircraft, 1969; classification "Secret—No Foreign Dissemination").

James, Peter N., *Annual Report on Foreign Rocket Technology (1969–1970)*, PWA-FR-3760 (West Palm Beach, Florida: Pratt & Whitney Aircraft, July 15, 1970; classification "Secret—No Foreign Dissemination").

James, Peter N., *Annual Report on Foreign Rocket Technology (1969–1970)*, PWA-FR-3760A (Pratt & Whitney Aircraft, Florida, September 1, 1970; Secret—No Foreign Dissemination).

James, Peter N., Personal Notes (1965–1973).

Khrushchev, Nikita, *Khrushchev Remembers*, Introduction, commentary and notes by Edward Crankshaw and translated and edited by Strobe Talbott, (Boston: Little, Brown and Company, 1970).

Kirchner, Walther, *History of Russia* (New York: Barnes & Noble, Inc., 1966).

Korol, Alexander G., *Soviet Research and Development* (Boston: The M.I.T. Press, 1965).

Laird, Melvin R., *A Statement by Secretary of Defense Melvin R. Laird: Fiscal Year 1971 Defense Program and Budget* (before the House Subcommittee on Department of Defense Appropriations), Department of Defense Report (Washington: U.S. Government Printing Office, 1970).

Laird, Melvin R., *Statement of Secretary of Defense Melvin R. Laird on the Fiscal Year 1972–76 Defense Program and the 1972 Defense Budget* (before the Senate Armed Services Committee), Department of Defense Report (Washington: U.S. Government Printing Office, 1971).

Laird, Melvin R., *Statement of Secretary of Defense Melvin R. Laird on the FY 1973 Defense Budget and FY 1973–1977 Program* (before the Senate Armed Services Committee), Department of Defense Report (Washington: U.S. Government Printing Office, 1971).

Lebedev, Vladimir and Yuri Gagarin, *Survival in Space* (New York: Frederick A. Praeger, September 1969).

Lovell, Stanley, *Of Spies and Stratagems* (New York: Prentice-Hall, 1963).

Lyons, Eugene, *Workers' Paradise Lost* (New York: Paperback Library, 1967).

McNamara, Robert S., *Statement of Secretary of Defense Robert S. McNamara on the Fiscal Year 1968–72 Defense Program and 1968 Defense Report* (before the House Armed Services Committee), Department of Defense Report (Washington: U.S. Government Printing Office, 1967).

Mehnert, Klaus, *Peking and Moscow* (New York: G. P. Putnam's Sons, 1963).

Miller, William J., *The Meaning of Communism* (Morristown, N.J.: Silver Burdette Company, a division of General Learning Corporation, 1968).

Nixon, President Richard M., *U.S. Foreign Policy for the 1970s, The Emerging Structure of Peace, A Report to the Congress* (Washington, U.S. Government Printing Office, 1972).

Penkovskiy, Oleg, *The Penkovskiy Papers*, translated by Peter Deryabin (New York: Doubleday, 1965).

Peters, Charles and Taylor Branch, *Blowing The Whistle* (New York: Praeger Publishers, 1972).

Pipes, Richard, *International Negotiation, Some Operational Principles of Soviet Foreign Policy* (Washington U.S. Government Printing Office, 1972).

Philby, Kim, *My Silent War* (New York: Grove Press, 1968).

Powers, Francis Gary with Curt Gentry, *Operation Overflight* (New York: Holt, Rinehart and Winston, 1970).

Proxmire, Senator William, *Report From Wasteland* (New York: Praeger Publishers, 1970).

Riabchikov, Evgeny, *Russians in Space* (New York: Doubleday, 1971).

Salisbury, Harrison E., *Russia* (New York: Atheneum—The New York Times Company, 1965).

Scheer, Robert, *The Diary of Che Guevara* (New York: Bantam, 1968).

Schlafly, Phyllis, *Safe—Not Sorry* (Alton, Illinois: Pere Marquette Press, 1967).

Schlafly, Phyllis and Chester Ward, *The Gravediggers* (Alton, Illinois: Pere Marquette Press, 1964).

Seth, Ronald, *The Executioners* (New York: Hawthorne Books, 1967).

Sokolovsky, V. D., *Military Strategy* (Moscow: Moscow Publishing House, 1968).

Taylor, John W. R., *Jane's All The World's Aircraft 1970–1971* (London: Jane's Yearbooks, Sampson Low, Marston & Company, Ltd., 1970–1971).

Toledano, Ralph de, *Spies, Dupes & Diplomats* (New Rochelle, N.Y.: Arlington House, 1967).

Tully, Andrew, *CIA—The Inside Story* (New York: William Morrow & Company, 1962).

Tully, Andrew, *The Super Spies* (New York: William Morrow & Co., 1969).

Udell, Gilman G., *Laws Relating to Espionage, Sabotage, Etc.* (Washington U.S. Government Printing Office, 1971).

Wise, David and Thomas B. Ross, *The Espionage Establishment* (New York: Random House, 1967).

The Acquisition of Weapons Systems, Hearings, Subcommittee on Economy in Government of the Joint Economic Committee, Congress of the United States, 91st Congress, Part 1 (Washington: U.S. Government Printing Office, 1971).

Aeronautics and Space Report of the President, transmitted to Congress January 1971 (Washington U.S. Government Printing Office, 1971).

The Budget of the United States Government, Fiscal Year 1972 (Washington: U.S. Government Printing Office, 1972).

Communist Global Subversion and American Security Volume I, printed for the use of the Senate Committee on the Judiciary (Washington: U.S. Government Printing Office, 1972).

Department of Defense Annual Report for Fiscal Year 1965 (Washington: U.S. Government Printing Office, 1967).

The Economics of Defense Spending, A Look at the Realities, July 1972, Department of Defense Comptroller (Washington: U.S. Government Printing Office, 1972).

Economic Report of the President, transmitted to Congress January 1972 (Washington: U.S. Government Printing Office, 1972).

How Foreign Policy is Made, U.S. Department of State (Washington: U.S. Government Printing Office, 1971).

"Laboratories in Orbit," by Y. Zaitsev, in *Soviet Report*, Center for Foreign Technology, Pasadena, California (No. 9, 1969).

Manned Space Flight U.S.-Soviet Rendezvous and Docking, Hearing, Subcommittee on Manned Space Flight, U.S. House of Representatives Committee on Science and Astronautics (Washington: U.S. Government Printing Office, 1972).

Military Implications of the Treaty on the Limitations of Anti-Ballistic Missile Systems and the Interim Agreement on Limitation of Strategic Offensive Arms, Hearing, Committee on Armed Services, U.S. Senate, 92nd Congress (Washington: U.S. Government Printing Office, 1972).

Report to the President and the Secretary of Defense on the Department of Defense, by the Blue Ribbon Defense Panel (Washington: U.S. Government Printing Office, 1970).

Review of the Soviet Space Program, by the Science Policy Research Division, Legislative Reference Service, Library of Congress (Washington: Government Printing Office, 1967).

Senate Hearings, Department of Defense Appropriations, 91st Congress, Fiscal Year 1971, Parts 1, 2, 3, and 4 (Washington: U.S. Government Printing Office, 1970).

Senate Hearings, Department of Defense Appropriations, 92nd Congress, First Session Fiscal Year 1972, Parts 1, 2, 3, and 4 (Washington: U.S. Government Printing Office, 1971).

Senate Hearings, Department of Housing and Urban Development; Space, Science, Veterans, and Certain other Independent Agencies Appropri-

ations, 92nd Congress, Fiscal Year 1972 (Washington: U.S. Government Printing Office, 1971).

Soviet Naval Activities in Cuba, Hearings before the Subcommittee on Inter-American Affairs, 91st Congress—Part 1 and 2 (Washington: Government Printing Office, 1971).

Soviet Space Program, 1962–1965; Goals and Purposes, Achievements, Plans, and International Implications (for the use of the Senate Committee on Aeronautical and Space Sciences), Legislative Reference Service, Library of Congress (Washington: U.S. Government Printing Office, 1966).

Soviet Space Programs, 1966–1970 (for the use of the Senate Committee on Aeronautical and Space Sciences), Congressional Research Service, Library of Congress (Washington: U.S. Government Printing Office, 1971).

Soviet Space Programs, 1971: A Supplement to the Corresponding Report Covering the Period 1966–1970 (for the use of the Senate Committee on Aeronautical and Space Sciences), Congressional Research Service, Library of Congress (Washington: U.S. Government Printing Office, 1972).

United States Government Organization Manual 1972/1973 (Washington; U.S. Government Printing Office, 1972).

United States Relations with Europe in the Decade of the 1970s, Hearings, Subcommittee on Europe of the Committee on Foreign Affairs House of Representatives, 91st Congress (Washington: U.S. Government Printing Office, 1970).

United States and Soviet Rivalry in Space: Who is Ahead, and How Do the Contenders Compare? by Charles S. Sheldon, II, Science Policy Research Division (Washington: Library of Congress Legislative Reference Service, 1969).

The Russian Embassy Information Department, Washington, D. C.

Subjective Analysis
Summary (SAS) Index

252

Soyuz, 58, 94, 102, 114-117, 145,
152-153, 155-156, 157 (photo),
162-163, 186, 202; *also see*
Orbit-to-orbit reusable shuttle.
United States
Apollo, 58, 94, 114-115, 117,
152-153, 156, 163, 186; *also see*
Apollo program, Moon race.
Gemini, 65, 94, 109, 113, 152, 156
Mercury, 65, 94, 113, 134, 152
Spacecraft (unmanned orbital): *Refer
to* Chapter 8; 109-112; *also see*
Orbital weapons, Satellite killer.
Spook (spy), 38
Sputnik, 41, 83, 94, 98, 129, 197
Spy-in-the-sky (satellite), 120
State-of-the-art, 57, 75, 88, 128
Strategic balance (U.S.-U.S.S.R.), 185
Strategic delivery vehicles (U.S.-
U.S.S.R.), 174
Strategic threat (U.S.S.R.): *Refer to*
Chapter 14; 167-182; *also see*
Bombers, ICBMs, SLBMs.
Submarines, 119, 178, 203; *also see*
SLBMs.
Surveyor (U.S. unmanned moon land-
ing craft), 89, 97

Tass (Soviet news agency), 39, 73, 159
TU-144 (Soviet passenger supersonic
transport), 73, 179, 201
Technological breakthrough, 70
Technological Pearl Harbor, 70, 191,
197, 204
Technology gap, 69
Trident (U.S. nuclear submarine), 203

U-2 (spy plane), 38, 122
Ukrainian, 36
United Nations, 38-39, 56, 122
United States Government
Air Force (USAF), 45, 47, 50-51,
72-73, 77, 85, 98, 118, 127-129,
143-144, 147-150, 156, 162,
199, 202-203

Aero Propulsion Laboratory (APL),
53
Aeronautical Systems Division
(ASD), 144; *also see* Fighter air-
craft (F-15 air superiority
fighter).
Arnold Engineering Development
Center (AEDC), 76
Edwards Air Force Base (Califor-
nia), 129
Foreign Technology Division
(FTD), 31, 43, 45, 57
Kirtland Air Force Base (New
Mexico), 188
Patrick Air Force Base (Florida),
46
Rocket Propulsion Laboratory
(RPL), 148-149; *also see* Rocket
engines (U.S. high chamber pres-
sure rocket engine).
Space and Missiles Systems Organi-
zation (SAMSO), 51, 53, 144;
also see Orbit-to-orbit reusable
shuttle (U.S.A.F.)
Strategic Air Command (SAC), 65,
121, 203
Systems Command, 177
Wright-Patterson Air Force Base
(Ohio), 53; *also see* Air Force
Foreign Technology Division.
Army, 50, 53, 150
Corps of Engineers, 46
CIA (Central Intelligence Agency),
31, 37-38, 44, 47, 49, 57, 70,
81, 87-88, 101, 103, 114, 131,
135, 139, 142, 193
Congress, 67, 129, 138-139, 171,
177, 187, 194-195, 198-199,
201
House of Representatives Armed
Services Committee, 29, 177-178
The Senate, 119, 170
Defense Science Board, 51
Department of Defense (DOD), 45-
48, 70, 73, 78, 114, 120, 129,